Surface Treatment of Textiles

Surface Treatment of Textiles

Editor

Chi-wai Kan

MDPI • Basel • Beijing • Wuhan • Barcelona • Belgrade • Manchester • Tokyo • Cluj • Tianjin

Editor
Chi-wai Kan
The Hong Kong Polytechnic University
Hong Kong

Editorial Office
MDPI
St. Alban-Anlage 66
4052 Basel, Switzerland

This is a reprint of articles from the Special Issue published online in the open access journal *Coatings* (ISSN 2079-6412) (available at: https://www.mdpi.com/journal/coatings/special issues/ Surf Text).

For citation purposes, cite each article independently as indicated on the article page online and as indicated below:

LastName, A.A.; LastName, B.B.; LastName, C.C. Article Title. *Journal Name* **Year**, *Volume Number*, Page Range.

ISBN 978-3-0365-2081-0 (Hbk)
ISBN 978-3-0365-2082-7 (PDF)

Contents

About the Editor

Chi-wai Kan is a professor at the Institute of Textiles and Clothing, The Hong Kong Polytechnic University.

Dr. Kan obtained a B.Sc. in textile chemistry and later a Ph.D. from The Hong Kong Polytechnic University. Before joining the Institute, he worked in private and public sectors in textile evaluation and safety and health management for more than 5 years. His main duties at the Institute are teaching coloration and finishing. D r. Kan's research interests are also in the area of (i) textile coloration and finishing, (ii) surface treatment of textile materials, (iii) textile products evaluation, (iv) textile testing instrumentation, (v) safety and health management, and (vi) environmental management.

Dr. Kan holds the professional qualifications of Chartered Colourist, Chartered Textile Technologist, Chartered Chemist, and Chartered Safety and Health Practitioner, and fellowships with the Society of Dyers and Colourists, the Textile Institute, and the Royal Society of Chemistry, U.K. In addition, Dr. Kan is a member of the Institution of Occupational Safety and Health and the Hong Kong Institution of Textile and Apparel.

Recently, Dr. Kan published more than 180 SCI journal papers and 250 conference papers in his research area.

Preface to "Surface Treatment of Textiles"

Textiles are commonly composed of natural and synthetic fibers for normal applications. To impart functional or aesthetic effects on textiles, the surface characteristics in the fiber play an important role. Therefore, surface treatment or modification is a possible way to provide value-added properties to textiles. The textile material/fiber surface can be treated or modified physically and/or chemically to achieve different desired effects.

In connection with different developments in the surface treatments of textiles with various applications, we developed a Special Issue in Coatings (https://www.mdpi.com/journal/coatings/special_issues/Surf_Text, accessed on 18 August 2021) that aims to provide an open forum to draw the attention of academic researchers and industrial experts to investigate different aspects of the surface treatment of textiles. This Special Issue contains five manuscripts that cover different aspects of surface treatment of textiles:

(1) Development of Durable Antibacterial Textile Fabrics for Potential Application in Healthcare Environment;

(2) Microstructure and Properties of Polytetrafluoroethylene Composites Modified by Carbon Materials and Aramid Fibers;

(3) Analysis of Surface Properties of Ag and Ti Ion-Treated Medical Textiles by Metal Vapor Vacuum Arc Ion Implantation;

(4) An Investigation on the Thermal and Solar Properties of Graphene-Coated Polyester Fabrics;

(5) Development of Advanced Textile Finishes Using Nano-Emulsions from Herbal Extracts for Organic Cotton Fabrics.

These manuscripts reflect the recent development of surface treatment of textiles with different applications. To introduce these recent developments to a wider readership, we collected these manuscripts and published this book. We hope this book will promote new research topics in the surface treatment of textiles in the academic field and help the industry look for new product ideas.

Chi-wai Kan
Editor

coatings

MDPI

Editorial

Special Issue "Surface Treatment of Textiles"

Chi-Wai Kan

Institute of Textiles and Clothing, The Hong Kong Polytechnic University, Hung Hom, Kowloon, Hong Kong, China; kan.chi.wai@polyu.edu.hk

Textiles are commonly composed of natural and synthetic fibers for normal applications. To impart functional or aesthetic effects on textiles, the surface characteristics in the fiber play an important role. Therefore, surface treatment or modification is a possible way to provide value-added properties to textiles. The textile material/fiber surface can be treated or modified physically and/or chemically to achieve different desired effects.

In connection with different development in the surface treatments of textiles with various applications, we developed a Special Issue in Coatings (https://www.mdpi.com/journal/coatings/special_issues/Surf_Text, accessed on 18 August 2021) which aims to provide an open forum to draw the attention of academic researchers and industrial experts to investigate different aspects of the surface treatment of textiles. This Special Issue contains five manuscripts that cover different aspects of surface treatment of textiles:

(1) Development of Durable Antibacterial Textile Fabrics for Potential Application in Healthcare Environment [1]
(2) Microstructure and Properties of Polytetrafluoroethylene Composites Modified by Carbon Materials and Aramid Fibers [2]
(3) Analysis of Surface Properties of Ag and Ti Ion-Treated Medical Textiles by Metal Vapor Vacuum Arc Ion Implantation [3]
(4) An Investigation on the Thermal and Solar Properties of Graphene-Coated Polyester Fabrics [4]
(5) Development of Advanced Textile Finishes Using Nano-Emulsions from Herbal Extracts for Organic Cotton Fabrics [5]

These manuscripts reflect the recent development of surface treatment of textiles with different applications. To introduce these recent developments to a wider readership, we collected these manuscripts and published this book. We hope that this book will promote new research topics in surface treatment of textiles in the academic field as well as help the industry to look for new production ideas.

Citation: Kan, C.-W. Special Issue "Surface Treatment of Textiles". *Coatings* **2021**, *11*, 984. https://doi.org/10.3390/coatings11080984

Received: 18 August 2021
Accepted: 18 August 2021
Published: 19 August 2021

Funding: This research received no external funding.

Institutional Review Board Statement: Not applicable.

Informed Consent Statement: Not applicable.

Data Availability Statement: Not applicable.

Acknowledgments: We would like to thank the Journal's editorial team in supporting the Special Issue as well as developing this book.

Conflicts of Interest: The authors declare no conflict of interest.

References

1. Wang, W.Y.; Chiou, J.C.; Yip, J.; Yung, K.F.; Kan, C.W. Development of durable antibacterial textile fabrics for potential application in healthcare environment. *Coatings* **2020**, *10*, 520. [CrossRef]
2. Zhang, F.; Zhang, J.; Zhu, Y.; Wang, X.; Jin, Y. Microstructure and properties of polytetrafluoroethylene composites modified by carbon materials and aramid fibers. *Coatings* **2020**, *10*, 1103. [CrossRef]
3. Akpek, A. Analysis of surface properties of Ag and Ti ion-treated medical textiles by metal vapor vacuum arc ion implantation. *Coatings* **2021**, *11*, 102. [CrossRef]
4. Manasoglu, G.; Celen, R.; Kanik, M.; Ulcay, Y. An investigation on the thermal and solar properties of graphene-coated polyester fabrics. *Coatings* **2021**, *11*, 125. [CrossRef]
5. Venkatraman, P.D.; Sayed, U.; Parte, S.; Korgaonkar, S. Development of advanced Textile finishes using nano-emulsions from herbal extracts for organic cotton fabrics. *Coatings* **2021**, *11*, 939. [CrossRef]

coatings

Article

Development of Durable Antibacterial Textile Fabrics for Potential Application in Healthcare Environment

Wen-Yi Wang [1], Jia-Chi Chiou [2], Joanne Yip [1], Ka-Fu Yung [2] and Chi-Wai Kan [1,*]

1 Institute of Textiles and Clothing, The Hong Kong Polytechnic University, Hung Hom, Kowloon, Hong Kong, China; tcwang@polyu.edu.hk (W.-Y.W.); joanne.yip@polyu.edu.hk (J.Y.)
2 Department of Applied Biology and Chemical Technology, The Hong Kong Polytechnic University, Hung Hom, Kowloon, Hong Kong, China; bcamber@polyu.edu.hk (J.-C.C.); bckfyung@polyu.edu.hk (K.-F.Y.)
* Correspondence: tccwk@polyu.edu.hk; Tel.: +852-2766-6531

Received: 17 April 2020; Accepted: 27 May 2020; Published: 29 May 2020

Abstract: Recently, efforts at development of functional textiles with antibacterial effect have accelerated, the purpose being to provide protection against airborne bio-particles and micro-organisms. Growth of microbes on surface of textile materials can be inhibited by biocidal approach and biostatic approach. This paper describes the development of a healthcare textile with durable antibacterial properties by optimizing the conventional and commercialized antimicrobial agent polybiguanide derivative—poly(hexamethylenebiguanide) (PHMB). Pad-dry-cure method was used to coat PHMB on cotton fabrics. The durability to simulated healthcare laundering of the fabric samples was evaluated in detail. Specifically, effects of detergent and washing cycles were examined. It was found that the optimum finishing condition can impart to the fabrics excellent durability for simulated healthcare washing. The fabric samples showed 100% bactericidal effect after 52 washing cycles, and 104 washings slightly reduced the bactericidal activity. Nevertheless, both simulated healthcare washing and coating treatment were found to have slightly negative influence on the hand feel and tearing strength properties of cotton fabrics.

Keywords: healthcare textile; antibacterial property; hand feel value; washing durability

1. Introduction

Cotton is a green natural cellulosic fiber available abundantly. It has been widely used in daily life because of its favorable properties that include high strength, soft handle, affinity to skin, and bio-degradability [1,2]. However, due to the morphology and moisture affinity of natural origin, cotton fabrics provide a perfect environment for growth of microorganism on the surface [3]. Depending on moisture, nutrients, temperature and pH value, the bacterial growth can be very fast, which may cause some undesirable effects, such as unpleasant odor, reduction of mechanical strength, stains, and discoloration [1,4]. Even mild bacterial growth can lead to rotting and breakdown of the fibers, which may make the fabric unattractive due to the appearance of unwanted pigmentation. Moreover, the microorganism's growth can accelerate the hydrolysis of cellulose and cause deterioration of the fabric.

Microorganisms that frequently contaminate surface of cotton fabrics include *Escherichia coli, Klebsiella pneumoniae, Pseudomonas aeruginosa, Staphylococcus aureus*, and *Acinetobacter baumannii*. These can cause pathogenic effects on human beings due to user contamination and cross infection [5,6]. Therefore, it is of great concern for the textile industry to develop antibacterial textiles. The demand for antibacterial textiles has been gaining increasing attention in the past few decades [7,8]. In healthcare environment, textiles with antibacterial properties form a significant part of the hygienic regime of surgical procedures, preventing the spread of infectious pathogens to both patients and staff.

Antibacterial finishing is thus typically carried out to give textiles improved resilience against microorganisms to prevent destruction of fibers and discoloration, and increased durability of the textiles with longer life, which plays an important role in addressing hygiene in clinical and sensitive environments by minimizing microbial colonization of textiles and the potential for transfer from fabric surfaces [9–11].

In this context, numerous methods have been developed to impart antibacterial properties to textiles to reduce the spread of microorganisms by killing or inhibiting their growth on the fabric surface, e.g., surface modification and finishing [3,12]. Gao and Cranston reviewed the application of various antibacterial agents to impart textiles with antimicrobial functions [3]. For examples, the conventional exhaust, pad-dry-cure processes, spraying and foam finishing method have been extensively studied for the antimicrobial finishing on textiles with triclosan and silicon-based quaternary agents [13,14]. Meanwhile, use of a wide variety of antibacterial agents in fabric finishing has been widely investigated, ranging from synthetic organic compounds, such as quaternary ammonium salt [9,15], polybiguanides [16,17], and *N*-halamines [18], through to TiO_2, ZnO, and silver nanoparticles [6,10,19] and naturally derived antimicrobials, such as chitosan [20]. Recently, the combination of antibacterial and hydrophobicity for cotton fabric has attracted much interest for better antibacterial behavior. For example, Mohammad and colleagues developed such a dual-functional antibacterial textile by immobilizing silver nanoparticles as the antibacterial agent and then modified it with octyltriethoxysilane (OTES) to achieve the hydrophobic surface [21]. Thirumalaisamy et al. explored the method of bimetallic deposition to obtain superhydrophobic and antibacterial cotton fabric [22]. Recently, some novel methods of antibacterial finishing have been reported. In this regard, Poonam and coworkers fabricated a dual-functional cotton fabric with superhydrophobicity and antibacterial property by simply immersing cotton fabric in non-fluorinated hexadecyltrimethoxysilane solution [23]. The modified cotton fabric not only shows excellent antibacterial property, but also has outstanding oil-water separation efficiency. Another example is the environmentally friendly antibacterial cotton fabric reported by Chen et al. [24], which was finished with siloxane sulfopropylbetaine (SSPB). SSPB was found not to be leachable from the fabric and the coated fabric is nontoxic to skin. Rong et al. reported the method of heterogeneous transesterification to develop hydrophobic and antibacterial cotton fabric via constructing enamine bonds between acetoacetyl groups and cotton fiber [1]. The resulting material showed excellent and durable bacteria-resistance rates of over 99.99% against *E. coli* and *S. aureus*. Additionally, novel graphene oxide based antibacterial cotton fabric has also been reported and extensively investigated by Zhao and his colleagues [25]. However, although these attempts are highly meaningful and of importance for academia, there are still a long way from large-scale industrial production.

Polyhexamethylene biguanide (PHMB) is a widely used polymeric antibacterial agent due to its low toxicity and broad-spectrum biocide property, which has been commercially available in the market [15,26]. It is usually used as a biocide, and is also an important drug used in several topical applications. PHMB is composed of repeated basic biguanidine units connected by hexamethylene hydrocarbon chains, providing a cationic and amphipathic structure (Figure 1). The antibacterial activity of PHMB depends on the molecular structure, which increases with the level of polymerization, and the maximum biocidal efficiency is obtained when the hexamethylene group is employed as the spacer [16]. Biguanides in the PHMB molecules account for the inhibition and binding activities of PHMB.

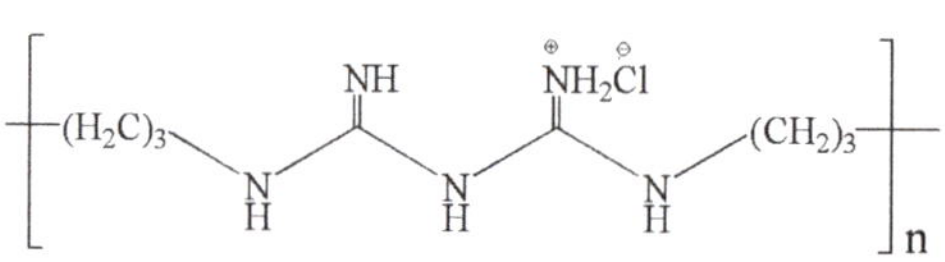

Figure 1. Chemical structure of poly(hexamethylenebiguanide) (PHMB).

However, a major concern for PHMB coating on fabric surface consists in the compromised tearing strength and hand feel values. The present study is thus focusing not only on the antibacterial behavior and durability, but also on the physical properties of cotton fiber after coating with PHMB formulation. The resulting cotton fabrics were subjected to detailed characterization by using a series of analytical techniques, including hand feel measurement and tearing strength.

2. Materials and Methods

2.1. Materials

One hundred percentwoven cotton fabric (fabric weight 272.8 g/cm^2) was used in this study. Originally, the fabric was scoured in an aqueous solution containing 5 g/L sodium hydroxide (chemical grade, obtained from Sigma-Aldrich, St. Louis, MO, USA) to remove impurities at temperature of 60 °C for 30 min, after which it was thoroughly rinsed with tap water and dried at room temperature.

Antibacterial agent PHMB (20% *w/v* aqueous solution), PEG400 (400 g/mol in average) and polyurethane binder (20% *w/v* aqueous solution) were supplied by Breakthrough Textiles Co., LTD. (Taipei, Taiwan). The commercial detergent (Attack, concentrated liquid, 3 L, Kao (HK) Corp., Tokyo, Japan) was obtained from a local supermarket. Non-ionic detergent, C13 Oxoalcoholethoxylate (7EO), was obtained from SDC Enterprises Limited, Holmfirth, UK. Sodium hypochlorite (NaClO) and bromophenol blue (BPB) sodium were obtained from Sigma-Aldrich (St. Louis, MO, USA).

2.2. Pad-Dry-Cure Treatment

Finishing formulations were prepared by adding different amounts of PHMB, PEG400, and binder to deionized water, and the "pad-dry-cure" method was used to coat the antibacterial agent on the cotton fabric. Specifically, the fabric sample was first padded with finishing formulations with 80% wet pickup and then dried in thermal oven (Werner Mathis AG, Switzerland) at 90 °C for 5 min and cured in Mathis LabDryer (Werner Mathis AG, Switzerland) at 130 °C for 45 s with fan speed of 1600 rpm. After antibacterial finishing, fabric samples were stored under standard conditions at 20 ± 2 temperature and 65 ± 2% relative humidity for at least 24 h prior to subsequent testing.

2.3. Washing Test

Washing test was conducted to evaluate effects of simulated healthcare laundering on antibacterial property of coated fabric samples. Different washing conditions with three washing cycles (26/52/104) were used (Table 1). The test was conducted in a laboratory scale washing machine (Launder Ometer, SDL Atlas, Rock Hill, SC, USA). After each washing cycle, the fabric sample was withdrawn and washed with tap water, and then the same volume of washing bath was added for the next cycle. Bleaching agent sodium hypochlorite (NaClO) was used in the washing solution.

Table 1. Washing test conditions [3].

	No Detergent 0	With Detergent 1
Solution 1 [1]	S10	S11
Solution 2 [2]	S20	S21

[1] Solution 1: 75 °C, 5 min, with or without detergent; [2] Solution 2: 65 °C, 10 min + NaClO (300 ppm), with or without detergent; [3] Washing cycles: 26/52/104.

2.4. Antibacterial Property

Antibacterial property of treated samples was qualitatively investigated against gram-positive *S. aureus* and gram-negative *K. pneumoniae* according to AATCC TM 147-2011, and the quantitative determination was performed in accordance with AATCC TM100-2019. The bacteria was grown in nutrient agar at 37 °C for 18 h. Fabric samples were cut into 20 mm × 20 mm pieces for the antibacterial test.

2.5. Scanning Electron Microscopy (SEM) Observation

Surface morphology of the fabric samples was observed by SEM (JEOL Model JSM-6490, JEOL USA, Inc., Peabody, MA, USA) with imaging up to 300,000× with a high resolution of 3 nm. The samples were treated by spray-gold prior to observation.

2.6. Qualitative Determination of PHMB by Dyeing with BPB

The anion of the aqueous sodium salt of BPB can be complexed with the cationic polymers on a substrate, leading to formation of a blue stable complex, which provides a standard method to determine cationic polymers [27,28]. The present study utilizes BPB to qualitatively determine the presence of PHMB coated on the fabrics.

2.7. Fourier Transform Infrared Spectroscopy Analysis

The physiochemical properties of PHMB-coated cotton fabric were analyzed via Fourier transform infrared spectroscopy (FTIR, Spectrum 100 FT-IR Spectrometer, Perkin Elmer, Waltham, MA, USA). The fabric samples were fastened on the sample stage and scanned from 400 cm^{-1} to 4000 cm^{-1} with a resolution of 4 cm^{-1}.

2.8. Tearing Strength Analysis

Tearing strength of fabric samples before and after PHMB coating was measured by Elmendorf Tear Strength Tester according to American Society for Testing and Materials (ASTM) D2261 standard. Three specimens (200 mm × 75 mm) were measured for each sample.

2.9. Hand Feel Measurement

The hand feel value was analyzed by the PhabrOmeter instrument (Nu Cybertec, Inc., Davis, CA, USA) according to AATCC Test Method 202-2014. For the purpose of discussion, hand feel index (HFI) is defined as summation of resilience, softness, and smoothness, as shown in Equation (1).

$$\text{HFI} = \text{Resilience} + \text{Softness} + \text{Smoothness}. \tag{1}$$

3. Results

3.1. Surface Morphology

Optical photographs and SEM were performed to observe the change in surface morphology of the fabric samples after coating with PHMB formulation. Clearly, the surface of cotton fabric was smooth (Figure 2A). From Figure 2B, the coated PHMB was obviously observed, and BPB dyeing confirms that antibacterial agent PHMB was successfully coated on the surface of the cotton fabric (Figure 2D).

3.2. FTIR Measuremnt

FTIR analysis was conducted to confirm the presence of PHMB on the treated cotton fabric, as shown in Figure 3. A sharp and strong absorbance peak at 1550 cm^{-1} was observed in the spectrum of PHMB, characteristic of imine group. The broad double peaks at 3195 cm^{-1} and 3380 cm^{-1} are ascribed to stretching vibration of –NH- and –NH_2- of PHMB molecule. Compared to PHMB, the absorption spectrum of cellulose is pretty weak, particularly for the untreated control cotton fabric. As shown in the insert, the characteristic peak of cellulose at 3330 cm^{-1} becomes stronger after coating with PHMB, which demonstrates that PHMB was successfully coated on the cotton fabric.

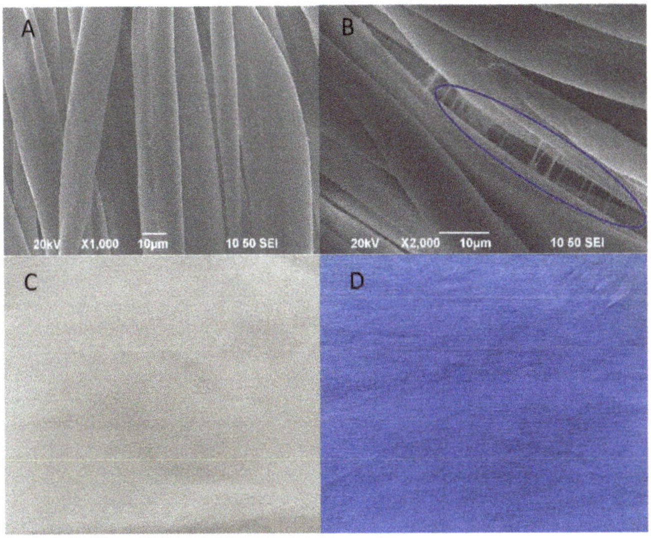

Figure 2. Surface morphology of treated fabric observed by SEM: (**A**) untreated cotton fabric, (**B**) fabric treated with PHMB, and optical photographs: (**C**) untreated cotton fabric and (**D**) PHMB-coated fabric dyed with bromophenol blue (BPB).

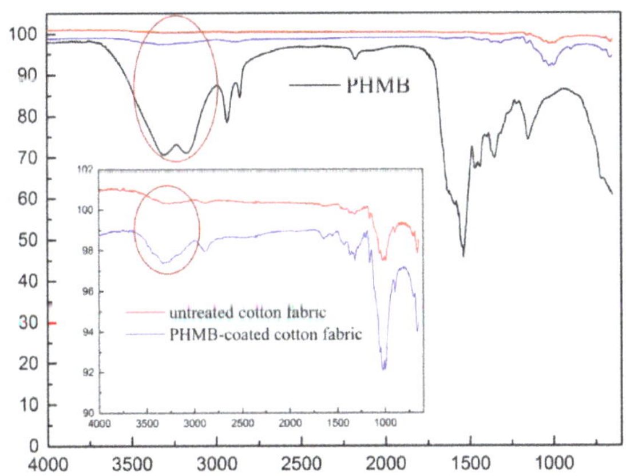

Figure 3. FTIR absorbance spectra of PHMB, untreated cotton fabric and PHMB-coated cotton fabric. The insert shows the spectrum of untreated cotton and treated counterpart.

3.3. *Effect of Concentration of PHMB on Antibacterial Activity*

In order to explore effects of PHMB on the antibacterial activity, different concentrations were further studied (Figure 4). It can be clearly seen that all the samples treated with different amounts of PHMB show strong inhibition against *S. aureus* and *K. pneumonia*. Table 2 shows that the concentration of PHMB had a significantly positive relationship with antibacterial activity. It is noteworthy that the antibacterial activity for concentration of 2.5 and 5% was not significant. A significant antibacterial ability was seen for concentrations of 10 and 15%, however.

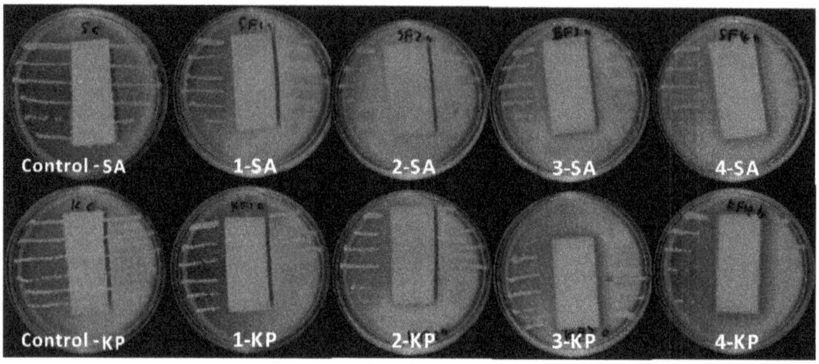

Figure 4. Qualitative assay for the antibacterial activity of PHMB-coated fabrics against *S. aureus* and *K. pneumonia*.

Table 2. Effect of concentration of PHMB on antibacterial activity.

Sample	PHMB Conc./%	Inhibition Zone/mm	
		S. aureus	*K. pneumoniae*
Control	0	0	0
1	2.5	4.5	3.5
2	5	4.5	5.5
3	10	4.75	6.75
4	15	6.75	7.25

3.4. *Effect of Detergent on Antibacterial Activity*

In practical application, medical textiles are inevitably subjected to simulated healthcare laundering. Frequent laundering generally has a serious negative influence on the antibacterial activity of the treated fabric. Therefore, durability against laundering is of high importance for medical textiles. This study is thus focused on the effect of healthcare laundering washing on the antibacterial ability. Firstly, we studied the effect of commercial detergent on antibacterial activity against *S. aureus* of treated fabric samples. The fabric sample used in this experiment was not treated with PHMB, but was just subjected to the washing test. Table 3 shows the results of samples washed by detergent with and without bleaching agent sodium hypochlorite (300 ppm). It was found that addition of commercial detergent in the washing solution shows strong inhibition and disinfection against *S. aureus*, and this effect is irrespective of the number of washing cycles. Moreover, the fabric samples showed yellowing after being washed with the addition of the bleaching agent sodium hypochlorite.

Table 3. Antibacterial properties of untreated samples washed by detergent.

Washing Cycles	Detergent		Detergent + NaClO	
	Inhibition/% [1]	Disinfection/% [2]	Inhibition/% [1]	Disinfection/% [2]
Control, no wash	0	0	0	0
26	0	100	0	100
52	0	100	0	100
104	0	100	0	100

[1] Inhibitory % = [(control after 18 h) − (specimen after 18 h)]/[control after 18 h] × 100%. [2] Disinfectant % = [(control after 0 h] − [specimen after 18 h)]/[control after 0 h] × 100%.

3.5. *Effect of Washing Cycles on Antibacterial Activity*

Table 4 shows the effects of washing cycles on the antibacterial activity against *S. aureus* and *K. pneumonia* of PHMB coated fabric samples. As shown in Table 4, after 52 washing cycles, the fabric samples still showed strong inhibition and disinfection, against both *S. aureus* and *K. pneumonia*. However, a clear decrease was observed in the antibacterial ability for the samples after 104 washing cycles. This indicates that simulated healthcare laundering has a significant influence on antibacterial activity of treated fabrics, and the coated samples show excellent durability against washing.

Table 4. Antibacterial properties of treated samples washed by different cycles.

Washing Cycles	Inhibition/% [1]		Disinfection/% [2]	
	S. aureus	*K. pneumoniae*	*S. aureus*	*K. pneumoniae*
no wash	100	100	100	100
26	100	100	100	100
52	100	100	100	100
104	95	38.6	0	0

[1] Inhibitory % = [(control after 18 h) − (specimen after 18 h)]/[control after 18 h] × 100%. [2] Disinfectant % = [(control after 0 h] − [specimen after 18 h)]/[control after 0 h] × 100%.

The qualitative analysis of PHMB-coated fabric samples was also conducted via dyeing with BPB. As seen in Figure 5, it clearly shows that the blue complex salt was significantly formed on the treated samples, which demonstrates that PHMB still existed on the fabric samples after different washing cycles.

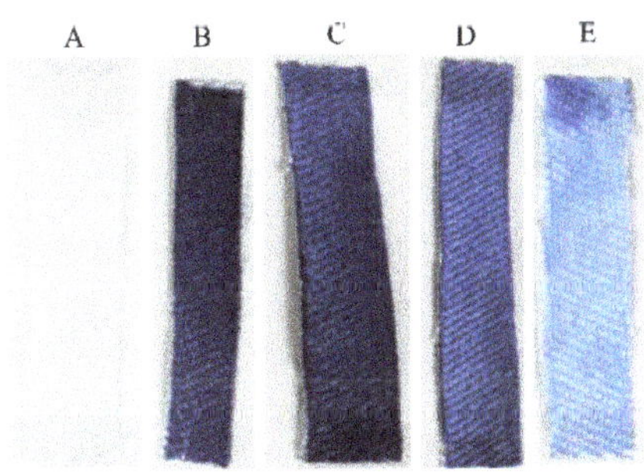

Figure 5. Qualitative test results of PHMB-coated fabric samples dyed with BPB: (**A**) pristine uncoated fabric; (**B**) no wash; (**C**) 26 washing cycles; (**D**) 52 washing cycles; and (**E**) 104 washing cycles.

3.6. Tearing Strength Analysis

Figure 6 presents that the effects of scouring, coating and washing on tearing strength of the fabric substrate. Clearly, for the control sample, tearing strength in both warp and weft directions was found to be the highest compared to the other samples. There is an obvious reduction in tearing strength of fabric samples after scouring, indicating that scouring has a negative effect on tearing strength of fabrics. Likewise, after coating the scoured fabrics, tearing strength in the warp direction dropped from 28.44 to 24.07, whereas the weft direction saw a decrease from 28.89 to 24.45. Then, the coated fabric sample was washed 104 times and tearing strength sharply declined to 14.7 and 13.86, for warp and weft direction, respectively. The results demonstrate that scouring, coating and washing treatment have a significant negative influence on tearing strength, particularly washing.

Figure 7 shows that the effect of washing conditions on tearing strength. It can be clearly seen that there was no significant difference between washing temperature 65 °C and 75 °C. Compared with simulated healthcare washing, tearing strength of treated fabrics has an obvious decrease, but it is still higher than that of washed samples. This demonstrates that the strength loss for treated samples is acceptable.

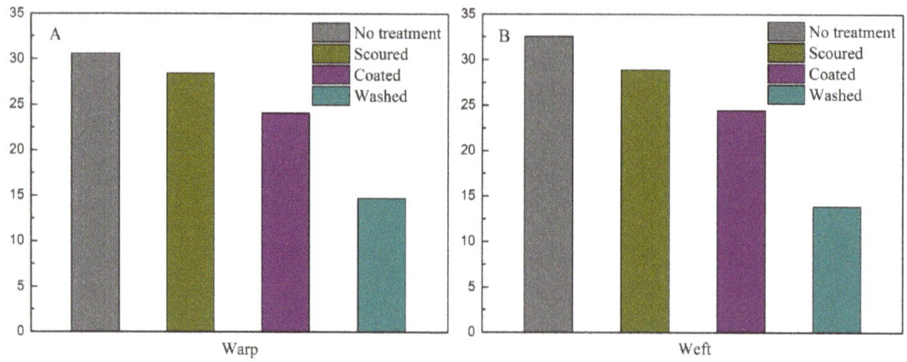

Figure 6. Tearing strength of control sample, scoured, coated, and washed samples (104 washing cycles, 65 °C with nonionic detergent but without bleaching agent); warp (**A**) and weft (**B**).

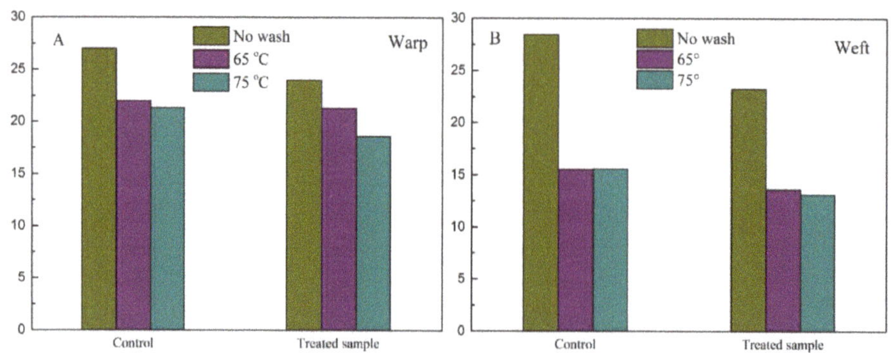

Figure 7. Tearing strength of control and treated samples after 52 washing cycles at 65 °C and 75 °C, warp.

3.7. Hand Feel Measurement

In the project, antibacterial finishing formulation was fully investigated to assess its effect on the hand feel property, as shown in Table 5.

Table 5. The different antibacterial recipes for fabric samples antibacterial treatment.

Sample	PHMB (%, *v*/*v*)	PEG (%, *v*/*v*)	Binder (%, *v*/*v*)
G1	5	5	5
G2	5	3	5
G3	5	1	5
G4	5	0	5
G5	5	5	3
G6	5	3	3
G7	5	1	3
G8	5	0	3

After coating with above recipes, the samples were then assessed in terms of hand feel properties, such as resilience, softness, and smoothness, etc., as shown in Table 6. Clearly, compared to the control sample, hand feel properties show a significant decrease in terms of resilience, softness, and smoothness. This indicates that the antibacterial treatment has a negative impact on the hand feel of fabrics. Likewise, an obvious decline in whiteness was also observed after coating with PHMB recipe. This demonstrates that antibacterial treatment has a slightly negative influence on the performance of fabrics.

Table 6. Hand feel properties of fabric samples coated with different antibacterial recipes.

Sample	Resilience	Softness	Smoothness	HFI	Whiteness
Control	82.3	78.4	82.3	243.0	84.6
G1	80.3	76.7	81.2	238.2	83.9
G2	82.4	77.3	81.2	240.9	83.5
G3	83.0	77.4	81.5	241.9	82.4
G4	82.9	78.1	81.3	242.3	84.2
G5	82.2	78.6	81.6	242.4	83.2
G6	82.3	78.4	81.2	241.9	84.2
G7	83.0	77.2	81.8	242.0	82.5
G8	82.9	77.4	81.9	242.2	83.2

Next, the effect of washing on performance of textile fabrics was further investigated (Table 7). Compared to Table 6, HFI values of all the samples significantly decreased after washing, which indicates that washing has an obvious influence on the hand feel of textiles. Meanwhile, the whiteness also saw a sharp decrease from 80+ to 60+, particularly for the coated samples. This may be due to the loss of fluorescent whitening agent after washing because they are often used for textiles to improve the whiteness [29,30].

Table 7. The hand feel properties of PHMB coated fabric samples after washing (104 cycles, 65 °C).

Sample	Resilience	Softness	Smoothness	HFI	Whiteness
Control	79.3	78.0	79.5	236.8	82.7
G1	80.2	76.4	80.1	236.7	61.1
G2	79.8	76.4	79.7	235.9	63.5
G3	80.5	76.4	80.4	237.3	63. 2
G4	80.3	76.6	78.0	234.9	61.9
G5	80.8	76.8	80.6	238.2	62.3
G6	79.6	77.2	79.2	236.0	61.0
G7	80.0	76.7	80.0	236.7	63.6
G8	81.3	76.9	80.4	238.6	61.9

4. Discussion

PHMB is a cationic biocide marketed worldwide due to its excellent antibacterial activity, chemical stability, low toxicity, and reasonable cost, and has been extensively utilized in textile industry [3,7]. However, a common issue facing PHMB-coated fabrics is the durability against household laundering. The present study explored the effects of various parameters, such as washing cycles and household detergents, on the antibacterial activity of PHMB-coated fabric samples, as well as effects of PHMB formulation on the physical properties of cotton substrate, including tearing strength and hand feel value.

From SEM observation, the presence of PHMB was confirmed, and the stable blue complex was formed with BPB dyebath (Figure 2). The characteristic absorption peak of cellulose was also enhanced due to the coating of PHMB (Figure 3). The PHMB-coated fabric samples exhibit prominent inhibition activity against *S. aureus* and *K. pneumonia* (Figure 4). Moreover, the antibacterial behavior of PHMB was shown to a concentration dependent effect. However, there is no significant difference between the concentrations of 2.5% and 5% PHMB in the antibacterial behavior, while, for 10% and 15% PHMB, a significant antibacterial activity was observed (Table 3). The effectiveness against cellular organisms is due to the positively charged biguanide group attached to a flexible spacer. PHMB binds to the negatively charged phosphate head groups of phospholipids at bacteria cell wall or virus envelope, causing increased rigidity, sinking non-polar segments into hydrophobic domains, disrupting the membrane with subsequent cytoplasmic shedding, and culminating in cell death [16].

Durability against household laundering is an important property for medical textiles. Originally, we performed the washing tests with the commercial household detergent added to simulate the practical application scenario. However, it was found that the antibacterial properties of washed samples were seriously interfered by the commercial detergent, the possible reason of which may be the addition of some additives with microbicidal efficacy (Table 3). Therefore, a non-ionic surfactant 7EO was used for further experiments. The bleaching agent sodium hypochloride was abandoned due to the fabrics yellowing caused.

Next, we performed the antibacterial experiments on treated fabrics against two bacteria: *S. aureus* and *K. pneumonia* to pinpoint the effect of simulated healthcare laundering on the antibacterial ability. The fabric samples were shown strong bactericidal effect against both bacteria after 52 washing cycles (Table 4). The attenuated bactericidal activity was still observed in the samples with 104 washing cycles, which was further confirmed by BPB dyeing tests (Figure 5). This shows that frequent washing may negatively influence the antibacterial ability of PHMB-coated fabrics. In spite of this, the washing experiments demonstrate that the fabric samples treated with PHMB formulation possess satisfactory durability against washing. Similar results were also reported by Chen and coworkers [31]. In their study, PHMB was padded on the fabric and then dried at 120 °C for 5 min. PHMB-coated fabric samples still showed excellent antimicrobial property against *S. aureus* (99%) and *K. pneumonia* (94%) after 25 washing cycles. Gao et al. coated PHMB on the wool fabric by exhaustion method and studied the antibacterial behaviors against gram-negative bacteria *E. coli* [32]. They found that PHMB-coated wool fabric was durable to 5 accelerated washing cycles (ISO 5A cycle, equivalent to 40–50 wool washing cycles) and maintained 99% bacterial reduction for 8% PHMB treated samples. In the present study, the durability against laundering was dramatically increased to almost 104 washing cycles by optimizing the finish process. On the one hand, the underlying reason may be the electrostatic attraction between PHMB and cellulosic molecules [33], which is enhanced by the addition of polyurethane binder and high-temperature curing. On the other hand, polyurethane binder acts as a cross-linker and form hydrogen bond with PHMB and cellulosic molecules, which eventually enhance the durability property of PHMB treated fabrics [34].

Tearing strength of fabric is affected by many parameters, such as fiber structure and post-finishing process. The present study investigates the effects of scouring, coating, and washing on tearing strength of the fabric (Figure 6). Compared to the control sample, tearing strength of the fabrics were clearly debased after treatment, particularly after washing. In this study, the samples were firstly subjected to scouring treatment to remove the impurities, which slightly decrease the tearing strength of fabrics. And then, the scoured samples were coated with PHMB formulation via pad-dry-cure process, which further deteriorate the tearing strength. Accordingly, the tearing strength of the coated fabrics were gravely undermined after 104 washing cycles, the potential reason of which may be the hydrolysis of cellulosic fibers [35,36]. However, the effect of washing temperature on the tearing strength was not significant, even though a much strength loss was observed with a higher washing temperature (75 °C) (Figure 7).

Finishing treatment and regular washing generally have an obvious negative influence on the hand feel of a fabric [37,38], as confirmed in Tables 6 and 7. For the samples of G1–G4, which were coated with different concentrations of PEG, there is an increase in the value of HFI, indicating that the reduction of PEG helps enhance the hand feel performance of fabrics. Looking at samples of G5–G8, concentration of the binder dropped from 5% to 3% and effects of PEG on the hand feel was further investigated. It can be found that HFI values increased, excluding the sample G5. Overall, HFI values of samples G5–G8 were higher than G1–G4 samples, which was due to the smaller amount of PHMB formulation used for the coating (Table 6). By contrast, HFI values of all the samples significantly decreased after washing, which indicates that washing has an obvious influence on the hand feel of textiles, and the whiteness also sharply decreased (Table 7).

In summary, the fabric samples coated with PHMB formulation show excellent antibacterial activity against *S. aureus* and *K. pneumonia*. Washing test shows that washing has a significantly negative influence on the antibacterial activity, tensile strength and hand values. However, the samples prepared by the optimum process exhibit satisfactory durability for simulated healthcare laundering.

5. Conclusions

Durability against simulated healthcare washing is an important property for medical textiles. The durability property can be achieved by optimizing the finishing process. It was found that the optimum finishing conditions can impart excellent durability to fabrics expected to undergo repeated simulated healthcare washing. After 104 washing cycles, the PHMB-coated fabric samples still show distinct inhibition activity, particularly against *S. aureus*. Additionally, the present study also shows that both simulated healthcare washing and coating treatment have negative influence on the hand feel behavior and tearing strength of cotton fabric. Therefore, the antibacterial finishing for textiles should not neglect the negative influence on the materials.

Author Contributions: Conceptualization, J.C., J.Y., K.-F.Y., and C.-W.K.; Data curation, W.-Y.W.; Formal analysis, W.-Y.W., J.-C.C., and C.-W.K.; Funding acquisition, C.-W.K.; Investigation, W.-Y.W.; Methodology, W.-Y.W., J.-C.C., and C.-W.K.; Project administration, C.-W.K.; Supervision, C.-W.K.; Validation, W.-Y.W.; Visualization, W.-Y.W.; Writing-original draft, W.-Y.W.; Writing-review & editing, C.-W.K. All authors have read and agreed to the published version of the manuscript.

Funding: This research was funded by The Hong Kong Polytechnic University, grant numbers ZVNM and ZVQJ.

Acknowledgments: The authors would like to thank the financial support from The Hong Kong Polytechnic University for this work under the project of account codes: ZVNM and ZVQJ.

Conflicts of Interest: The authors declare no conflict of interest.

References

1. Rong, L.; Liu, H.; Wang, B.; Mao, Z.; Xu, H.; Zhang, L.; Zhong, Y.; Feng, X.; Sui, X. Durable antibacterial and hydrophobic cotton fabrics utilizing enamine bonds. *Carbohydr. Polym.* **2019**, *211*, 173–180. [CrossRef] [PubMed]
2. Periolatto, M.; Ferrero, F.; Vineis, C.; Varesano, A.; Gozzelino, G. Novel Antimicrobial Agents and Processes for Textile Applications. In *Antibacterial Agents*; IntechOpen: London, UK, 2017; Chapter 2.
3. Yuan, G.; Cranston, R. Recent Advances in Antimicrobial Treatments of Textiles. *Text Res. J.* **2008**, *78*, 60–72. [CrossRef]
4. Yousefi, M.; Dadashpour, M.; Hejazi, M.; Hasanzadeh, M.; Behnam, B.; De La Guardia, M.; Shadjou, N.; Mokhtarzadeh, A. Anti-bacterial activity of graphene oxide as a new weapon nanomaterial to combat multidrug-resistance bacteria. *Mater. Sci. Eng. C* **2017**, *74*, 568–581. [CrossRef] [PubMed]
5. Huang, T.; Chen, C.; Li, D.; Ek, M. Hydrophobic and antibacterial textile fibres prepared by covalently attaching betulin to cellulose. *Cellulose* **2019**, *26*, 665–677. [CrossRef]
6. Dastjerdi, R.; Montazer, M. A review on the application of inorganic nano-structured materials in the modification of textiles: Focus on anti-microbial properties. *Colloid Surf. B* **2010**, *79*, 5–18. [CrossRef] [PubMed]
7. Morais, D.; Guedes, R.; Lopes, M.A. Antimicrobial Approaches for Textiles: From Research to Market. *Materials* **2016**, *9*, 498. [CrossRef] [PubMed]
8. Gargoubi, S.; Tolouei, R.; Chevallier, P.; Levesque, L.; Ladhari, N.; Boudokhane, C.; Mantovani, D. Enhancing the functionality of cotton fabric by physical and chemical pre-treatments: A comparative study. *Carbohyd. Polym.* **2016**, *147*, 28–36. [CrossRef]
9. Windler, L.; Height, M.; Nowack, B. Comparative evaluation of antimicrobials for textile applications. *Environ. Int.* **2013**, *53*, 62–73. [CrossRef]
10. Hebeish, A.; El-Naggar, M.; Fouda, M.M.; Ramadan, M.A.; Al-Deyab, S.S.; El-Rafie, M. Highly effective antibacterial textiles containing green synthesized silver nanoparticles. *Carbohydr. Polym.* **2011**, *86*, 936–940. [CrossRef]

11. Xue, Y.; Xiao, H.; Zhang, Y. Antimicrobial Polymeric Materials with Quaternary Ammonium and Phosphonium Salts. *Int. J. Mol. Sci.* **2015**, *16*, 3626–3655. [CrossRef]
12. Chen, S.; Yuan, L.; Li, Q.; Li, J.; Zhu, X.; Jiang, Y.; Sha, O.; Yang, X.; Xin, J.H.; Wang, J.; et al. Durable Antibacterial and Nonfouling Cotton Textiles with Enhanced Comfort via Zwitterionic Sulfopropylbetaine Coating. *Small* **2016**, *12*, 3516–3521. [CrossRef] [PubMed]
13. Zhang, S.; Yang, X.; Tang, B.; Yuan, L.; Wang, K.; Liu, X.; Zhu, X.; Li, J.; Ge, Z.; Chen, S.; et al. New insights into synergistic antimicrobial and antifouling cotton fabrics via dually finished with quaternary ammonium salt and zwitterionic sulfobetaine. *Chem. Eng. J.* **2018**, *336*, 123–132. [CrossRef]
14. Mohsin, M.; Sardar, S. Development of sustainable and cost efficient textile foam-finishing and its comparison with conventional padding. *Cellulose* **2020**, *27*, 4091–4107. [CrossRef]
15. Simončič, B.; Tomsic, B. Structures of Novel Antimicrobial Agents for Textiles-A Review. *Text. Res. J.* **2010**, *80*, 1721–1737. [CrossRef]
16. Chindera, K.; Mahato, M.; Sharma, A.K.; Horsley, H.; Kloc-Muniak, K.; Kamaruzzaman, N.F.; Kumar, S.; McFarlane, A.; Stach, J.; Bentin, T.; et al. The antimicrobial polymer PHMB enters cells and selectively condenses bacterial chromosomes. *Sci. Rep.* **2016**, *6*, 23121. [CrossRef] [PubMed]
17. Sanada, H.; Nakagami, G.; Takehara, K.; Goto, T.; Ishii, N.; Yoshida, S.; Ryu, M.; Tsunemi, Y. Antifungal Effect of Non-Woven Textiles Containing Polyhexamethylene Biguanide with Sophorolipid: A Potential Method for Tinea Pedis Prevention. *Healthcare* **2014**, *2*, 183–191. [CrossRef]
18. Dong, A.; Wang, Y.-J.; Gao, Y.; Gao, T.; Gao, G. Chemical Insights into AntibacterialN-Halamines. *Chem. Rev.* **2017**, *117*, 4806–4862. [CrossRef]
19. Hebeish, A.; El-Naggar, M.E.; Tawfik, S.; Zaghloul, S.; Sharaf, S. Hyperbranched polymer–silver nanohybrid induce super antibacterial activity and high performance to cotton fabric. *Cellulose* **2019**, *26*, 3543–3555. [CrossRef]
20. Tayel, A.A.; Moussa, S.H.; El-Tras, W.F.; Elguindy, N.; Opwis, K. Antimicrobial textile treated with chitosan from Aspergillus niger mycelial waste. *Int. J. Boil. Macromol.* **2011**, *49*, 241–245. [CrossRef] [PubMed]
21. Khalil-Abad, M.S.; Yazdanshenas, M.E. Superhydrophobic antibacterial cotton textiles. *J. Colloid Interface Sci.* **2010**, *351*, 293–298. [CrossRef] [PubMed]
22. Suryaprabha, T.; Sethuraman, M.G. Design of electrically conductive superhydrophobic antibacterial cotton fabric through hierarchical architecture using bimetallic deposition. *J. Alloy. Compd.* **2017**, *724*, 240–248.
23. Chauhan, P.; Kumar, A.; Bhushan, B. Self-cleaning, stain-resistant and anti-bacterial superhydrophobic cotton fabric prepared by simple immersion technique. *J. Colloid Interface Sci.* **2019**, *535*, 66–74. [CrossRef] [PubMed]
24. Chen, S.; Chen, S.; Jiang, S.; Xiong, M.; Luo, J.; Tang, J.; Ge, Z. Environmentally Friendly Antibacterial Cotton Textiles Finished with Siloxane Sulfopropylbetaine. *ACS Appl. Mater. Interfaces* **2011**, *3*, 1154–1162. [CrossRef] [PubMed]
25. Zhao, J.; Deng, B.; Lv, M.; Li, J.; Zhang, Y.; Jiang, H.; Peng, C.; Li, J.; Shi, J.; Huang, Q.; et al. Graphene Oxide-Based Antibacterial Cotton Fabrics. *Adv. Heal. Mater.* **2013**, *2*, 1259–1266. [CrossRef]
26. Llorens, E.; Calderon, S.; Del Valle, L.J.; Puiggalí, J. Polybiguanide (PHMB) loaded in PLA scaffolds displaying high hydrophobic, biocompatibility and antibacterial properties. *Mater. Sci. Eng. C* **2015**, *50*, 74–84.
27. ISO Central Secretariat. *EN ISO 2871-1:2010 Surface Active Agents—Detergents—Determination of Cationic-Active Matter Content—Part 1: High-Molecular-Mass Cationic-Active Matter*; ISO Central Secretariat: Geneva, Switzerland, 2010.
28. ISO Central Secretariat. *EN ISO 2871-2:2010 Surface Active Agents—Detergents—Determination of Cationic-Active Matter Content—Part 2: Cationic-Active Matter of Low Molecular Mass (between 200 and 500)*; ISO Central Secretariat: Geneva, Switzerland, 2010.
29. Wan, M.; Hua, L.; Zeng, Y.; Jiao, P.; Xie, D.; Tong, Z.; Wu, G.; Zhou, Y.; Tang, Q.; Mo, F. Synthesis and properties of novel stilbene-twelve alkyl quaternary ammonium salts as antibacterial optical whitening agents. *Cellulose* **2017**, *24*, 3209–3218. [CrossRef]
30. Bueno, L.; Amador, C.; Bakalis, S. Modeling the deposition of fluorescent whitening agents on cotton fabrics. *AIChE J.* **2017**, *64*, 1305–1316. [CrossRef]
31. Chen-Yu, J.H.; Eberhardt, D.M.; Kincade, D.H. Antibacterial and Laundering Properties of AMS and PHMB as Finishing Agents on Fabric for Health Care Workers' Uniforms. *Cloth. Text. Res. J.* **2007**, *25*, 258–272. [CrossRef]

32. Gao, Y.; Yu, X.; Pierlot, A.P.; Denning, R.; Cranston, R. A simultaneous antimicrobial and shrink resistance treatment of wool woven fabrics using the polymeric biocide polyhexamethylene biguanide. *J. Mater. Sci.* **2011**, *46*, 3020–3026. [CrossRef]
33. Blackburn, R.S.; Harvey, A.; Kettle, L.L.; Payne, J.D.; Russell, S.J. Sorption of Poly(hexamethylenebiguanide) on Cellulose: Mechanism of Binding and Molecular Recognition†. *Langmuir* **2006**, *22*, 5636–5644.
34. Lomax, G.R. Breathable polyurethane membranes for textile and related industries. *J. Mater. Chem.* **2007**, *17*, 27757. [CrossRef]
35. Sarkar, J.; Khalil, E. Effect of Industrial Bleach Wash and Softening on the Physical, Mechanical and Color Properties of Denim Garments. *IOSR J. Polym. Text. Eng.* **2014**, *1*, 46–49. [CrossRef]
36. Sarkar, J.; Khalil, E.; Solaiman, M. Effect of Enzyme Washing Combined with Pumice Stone on the Physical, Mechanical and Color Properties of Denim Garments. *Int. J. Res. Advent Technol.* **2014**, *2*, 65–68.
37. Halleb, N.A.; Sahnoun, M.; Cheikhrouhou, M. The effect of washing treatments on the sensory properties of denim fabric. *Text. Res. J.* **2014**, *85*, 150–159. [CrossRef]
38. Uren, N.; Okur, A. Analysis and improvement of tactile comfort and low-stress mechanical properties of denim fabrics. *Text. Res. J.* **2019**, *89*, 4842–4857. [CrossRef]

coatings

MDPI

Article

Microstructure and Properties of Polytetrafluoroethylene Composites Modified by Carbon Materials and Aramid Fibers

Fubao Zhang *, Jiaqiao Zhang, Yu Zhu, Xingxing Wang and Yuyang Jin

School of Mechanical Engineering, Nantong University, Nantong 226019, China; 1810310038@stmail.ntu.edu.cn (J.Z.); zhuyu.ntu@gmail.com (Y.Z.); wangxx@ntu.edu.cn (X.W.); jinyuyang.ntu@gmail.com (Y.J.)

* Correspondence: zhang.fb@ntu.edu.cn; Tel.: +86-13646288919

Received: 12 October 2020; Accepted: 16 November 2020; Published: 18 November 2020

check for updates

Abstract: Polytetrafluoroethylene (PTFE) is polymerized by tetrafluoroethylene, which has high corrosion resistance, self-lubrication and high temperature resistance. However, due to the large expansion coefficient, high temperature will gradually weaken the intermolecular bonding force of PTFE, which will lead to the enhancement of permeation absorption and the limitation of the application range of fluoroplastics. In order to improve the performance of PTFE, the modified polytetrafluoroethylene, filled by carbon materials and aramid fiber with different scales, is prepared through the compression and sintering. Moreover, the mechanical properties and wear resistance of the prepared composite materials are tested. In addition, the influence of different types of filler materials and contents on the properties of PTFE is studied. According to the experiment results, the addition of carbon fibers with different scales reduces the tensile and impact properties of the composite materials, but the elastic modulus and wear resistance are significantly improved. Among them, the wear rate of 7 μm carbon fiber modified PTFE has decreased by 70%, and the elastic modulus has increased by 70%. The addition of aramid fiber filler significantly reduces the tensile and impact properties of the composite, but its elastic modulus and wear resistance are significantly improved. Among them, the wear rate of the modified composite material with 3% alumina particles and 5% aramid pulp decreased by 68%, and the elastic modulus increased by 206%.

Keywords: polytetrafluoroethylene; carbon material; aramid fiber; material modification

1. Introduction

PTFE is polymerized by tetrafluoroethylene, and its molecular chain is -[–CF_2–CF_2–]-^{n}. In detail, the conformation of PTFE molecular chain presents a spiral structure, and the C–F bond is one of the strongest chemical bonds, making the main chain structure of PTFE very stable [1]. PTFE has incomparable corrosion resistance, self-lubrication and high and low temperature resistance, which are unmatched by metal materials [2]. Thus, it is often used on the surface of some objects, such as textiles [3,4], valves [5], pipes [6], etc., through the fluorine lining process. Especially when PTFE is used as a sealing material, it is necessary to increase the elastic modulus of PEFT to prevent creep phenomenon [7], which will lead to leakage. Therefore, how to improve the corrosion resistance and elastic modulus of PTFE and expand the application range of PTFE on the surface of engineering objects has become a research hotspot [8].

The filling is one of the simplest and most effective modification methods for fluoroplastics [9,10]. More specifically, by adding fillers to the fluoroplastic matrix, the crystal structure of the fluoroplastic is changed, thereby improving and overcoming the original defects of the fluoroplastic [11]. The materials filled with modified PTFE mainly include carbon materials and aramid fibers [12,13]. In the modification of PTFE-filled fiber, the fiber mainly plays the role of bearing, and the matrix material plays the role of bonding the fibers firmly to each other and transferring the load to the fibers. Carbon fiber (CF) is a new type of fiber material with the carbon content of more than 95% after carbonization and graphitization of organic fibers [14,15]. The CF has excellent corrosion resistance and high temperature resistance, which is widely used as a high-quality filler or performance improver for rubber, plastic and various composite materials. Multiwalled carbon nanotube (MWCNT) is a quantum material with a special structure, which has good mechanical properties and more stable structure than polymers [16]. It is commonly used as the enhancer to improve the strength, elasticity and fatigue resistance of composite materials. Kevlar poly-p-phenylene terephthamide (PPTA) is a synthetic fiber with ultra-high strength and abrasion resistance [17]. PPTA has a high elastic modulus and excellent dimensional stability, and is widely used in the filling modification of polymer materials [18]. Al_2O_3 is a kind of metal aluminum oxide [19–21], it is often used as an inorganic filler to improve the hardness and wear resistance of modified PTFE materials. Due to the lack of necessary affinity between the inorganic filler and the PTFE interface, the combination of the two is poor, resulting in a decrease in the tensile strength, elongation, impact performance and compression rate of the modified PTFE material.

Scholars have conducted many studies on modification experiments of PTFE, and certain properties of PTFE have been improved. A.P. Vasilev studied the effect of composite fillers on the properties and structure of PTFE, and confirmed that the introduction of composite fillers significantly improved the wear resistance of PTFE composites [22]. A.A. Ohlopkova studied the effect of carbon fiber on the performance and structure of polytetrafluoroethylene. It is proved that carbon fiber can improve the tribological properties of PTFE [23]. Prateek Saxena et al. used a variety of filler materials to improve the frictional mechanical properties of cyanate ester, and proved that graphite produced the best effect among all fillers used, offering a lower specific wear rate, a lower friction value and a higher tensile strength and tensile modulus [24]. In addition, Prateek Saxena's team has developed a new type of testing device to identify the friction characteristics of carbon fiber reinforced polymer composites under high surface pressure, which is conducive to the performance testing of polymer materials [25]. Fuchuan Luo et al. used $Na_{1/2}Sm_{1/2}TiO_3$ and glass fiber to fill PTFE to improve the electrical conductivity of PTFE composites [26]. YingYuan et al. prepared a kind of PTFE composite material filled with Si_3N_4. As the content of Si_3N_4 increases, the electrical conductivity and thermal conductivity of the PTFE composite material are improved [27]. L.J. van Rooyen et al. used graphene oxyfluoride to fill PTFE, thereby reducing the permeability of PTFE. The incorporation of the oxyfluorinated and commercial grade graphene into the PTFE reduced the helium gas permeability by 96% at 4 vol.% and by 88% at 7 vol.% nanofiller contents, respectively [28].

Compared with previous studies, this article innovatively uses carbon nanotubes and aramid fibers to fill modified PTFE, thereby improving the wear resistance and elastic modulus of the composite material, making it more suitable for engineering object surfaces and seals. Starting from the modification process of fluoroplastics, this paper studies the influence of different types of filler materials on the properties of PTFE, and analyzes the mechanism of the influence of filler content on the properties of PTFE. Our research has effectively improved the wear resistance and elastic modulus of PTFE materials, which is conducive to the use of PTFE on the surface and seals of engineering objects, and provides scientific basis and reference for the research and practical application of new PTFE composite materials.

2. Materials and Instruments

2.1. Experimental Materials

The main reagents and materials used in the experiment are shown in Table 1.

Table 1. Main reagents and materials used in the test.

Reagent Name	Reagent/Material Manufacturer	Description of Product Codes
PTFE	Japan Daikin Fluorochemical (China) Co., Ltd. (Changshu, China)	Particle size of 20 μm
7 μm carbon fiber	Chuangjia Welding Material Co., Ltd. (Qinghe, China)	Length of 50 μm, monofilament diameter of 7 μm
200 nm carbon fiber	Shanghai Alighting Biochemical Technology Co., Ltd. (Shanghai, China)	Length of 5–50 μm, outer diameter of 200–600 nm
Multi-walled carbon nanotubes	Shanghai McLean Biochemical Technology Co., Ltd. (Shanghai, China)	Carboxylation, length of 50 μm, inner diameter of 3–5 nm, outer diameter of 8–15 nm, –COOH content of 2.6 wt.%
Aramid Pulp	Dupont China Group Co., Ltd. (Shenzhen, China)	Kevlar 1F1710, length of 0.7–1.6 mm
Al_2O_3	Mingshan New Materials Co., Ltd. (Laiwu, China)	Powder, particle size of 30–40 nm
Titanate coupling agent	Ding Hai Plastic Chemical Co., Ltd. (Dongguan, China)	KH-792
Absolute ethanol	Sinopharm Chemical Reagent Co., Ltd. (Shanghai, China)	Analytical reagent
Acetic anhydride	Jiangsu Qiangsheng Functional Chemical Co., Ltd. (Changshu, China)	Analytical reagent
Acetone	Shanghai Lingfeng Chemical Reagent Co., Ltd. (Shanghai, China)	Analytical reagent

2.2. Experimental Instruments

The main performance test instruments used to complete all the tests are shown in Table 2.

Table 2. Main test and analysis instruments used in the experiment.

Device Name	Device Model	Vendor
Compressive and flexural testing machine	YAW-300C	Jinan Dongfang Experimental Instrument Co., Ltd. (Jinan, China)
Shore Hardness Tester	LX-D type	Nanjing Suce Measuring Instruments Co., Ltd. (Nanjing, China)
Scanning electron microscope	S-3400N	Hitachi Manufacturing Co., Ltd. (Tokyo, Janpan)
Transmission electron microscope	JEM-2100Hu	Japan Electronics Corporation (Tokyo, Japan)
Pendulum impact sample machine	E45.105	MTS Industrial Systems Co., Ltd. (Shenzhen, China)
Universal testing machine	CMT5105	MTS Industrial Systems (China) Co., Ltd. (Shenzhen, China)
Friction and wear tester	MFT-V	American Rtec Equipment Co., Ltd. (San Jose, CA, USA)

2.3. Performance Testing and Organization Analysis Methods

The hardness testing shall be conducted in accordance with GB/T2411-2008 [29] Plastics and Ebonite Determination of Indentation Hardness (Shore hardness) using a Durometer. The LX-D plastic shore hardness tester was used to measure the shore hardness of friction and wear samples, using a D type indenter with a total test force of 588.4 N. Each sample measured 5 points, and the results are represented by the arithmetic mean of the 5 measured values.

The tensile testing was carried out in accordance with GB/T1040-2006 [30] Determination of Tensile Properties of Plastics. According to the standard, the sample was processed into a rectangular spline of 80 mm × 10 mm × 4 mm, and then tested on the electronic tensile testing machine of CMT5504 model at a beam speed of 10 mm/min. Three samples were prepared for each test, and the final results were taken as the tensile strength, elastic modulus and elongation at break according to the average value of the three samples.

The impact testing shall be conducted in accordance with GB/T1043-2008 [31] Determination of Impact Performance of Plastic Simply Supported Beams. According to the standard, the sample was processed into 80 mm × 10 mm × 4 mm with a depth of 2 mm A-shaped notch, and then the notched sample impact strength was measured on the combined impact tester of ZBC3000 model. The maximum impact energy of the pendulum was 7.5 J. Three samples were prepared for each test, and the final result is the average value of the three samples to express its impact toughness.

The friction and wear test was performed on the friction and wear tester of MFT-V model. The complex friction test method was adopted. The specific steps were to fix the sample on the chassis and reciprocate with the chassis. The grinding parts were made of cylindrical Cr15 bearing steel with a diameter of 5 mm, the friction stroke was set to 20 mm, the acceleration time was set to 1 s, and the reciprocating frequency was set to 2 Hz. The experiment was carried out under the load of 98 N, and the wear time was 30 min.

Impact section and wear surface morphology were performed by the following steps: first, the cross-section of the impact sample and the surface wave of the friction and wear sample were sprayed with gold to increase the conductivity of the composite material. Then, S-3400N scanning electron microscope (SEM) was used to observe the impact section and wear surface morphology under high magnification.

3. Preparation of Samples

3.1. Preparation of Carbon Material Modified PTFE

3.1.1. Surface Treatment of Carbon Fiber

Carbon fiber and nano Al_2O_3 ceramic particles are inorganic particles, which have poor binding properties with PTFE. Therefore, it is necessary to modify the surface of carbon fiber by the titanate coupling agent [32,33]. More specifically, the titanate coupling agent is coupled by its alkoxy group directly with the trace carboxyl groups or hydroxyl groups adsorbed on the surface of the filler or pigment to couple, thereby effectively improving the bonding strength between carbon fiber and PTFE.

3.1.2. Preparation of Modified PTFE

First of all, the filling modification needs to solve the problem of the dispersibility of the filler in PTFE, because the dispersion of the filler in the PTFE matrix will directly affect the performance of the composite material.

In this paper, the dry–wet mixing method was used to mix the composite materials [34], and the SEM micrograph of the carbon fiber used is as shown in Figure 1. First, the pretreated PTFE powder and carbon materials were weighed according to the mass percentage. Then, a certain amount of ethanol solution was poured into it and ultrasonic dispersion was performed for 30 min. In the next step, the PTFE powder after being stirred by a high-speed mixer and sieved was slowly poured into the stirring ethanol solution. It should be noted that the solution was stirred at high speed under the condition of heating in a water bath until the mixture shows a jelly without stratification. It is necessary to prevent the delamination of the solution during mixing. Then, the solution was placed in a drying cabinet at 80 °C for drying. The dried mixture was crushed by a high-speed crusher for 5 min and then sieved with a 100-mesh sieve to further ensure the homogeneity of the mixture. Then, the powder was placed in a mold for cold pressing to obtain a preformed sample. Next, the preformed sample was sintered in a box-type electric furnace to obtain a composite sintered sample. Finally, the test sample was cut with a punching knife for testing. The modified PTFE test sample is shown in Figure 2.

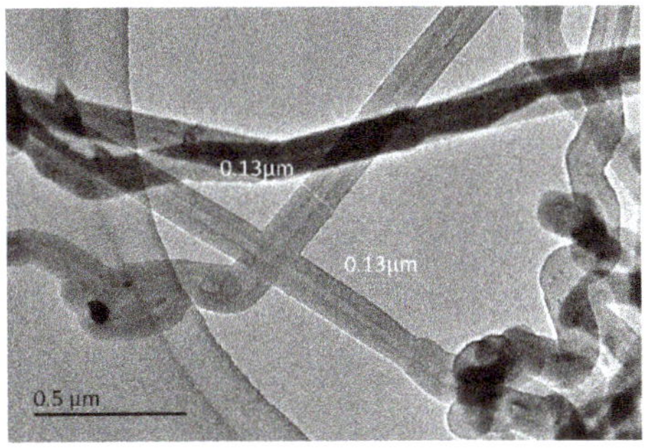

Figure 1. The SEM micrograph of the carbon fiber.

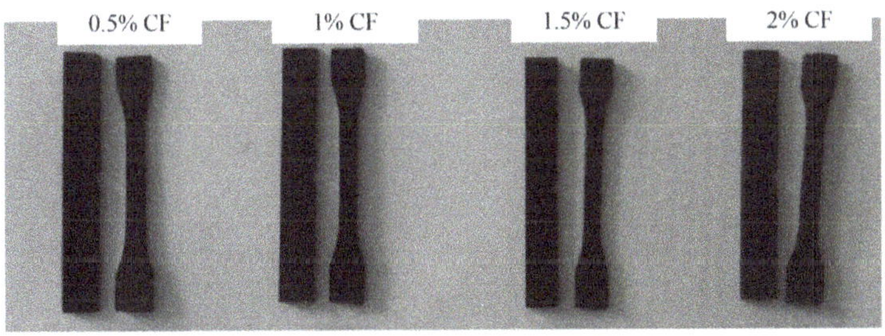

Figure 2. Test sample of carbon fiber (CF) modified PTFE composite.

3.2. Preparation of the Aramid Modified PTFE Material

3.2.1. Surface Treatment of the Aramid Fiber

The surface of the product often produces some defects during processing [35]. In order to ensure the cleanliness of the PPTA surface, it is necessary to clean the PPTA first. The detailed cleaning method is to soak PPTA in acetone for 1 h, then boil it with absolute ethanol and distilled water, and then filter the mixture and put it in an oven to dry for later use. The wastewater is treated in a harmless manner [36,37]. The SEM morphology of the PPTA surface is shown in Figure 3a. It can be seen that there are more attachments on the surface of PPTA. The SEM morphology of PPTA after cleaning with acetone is shown in Figure 3b. It can be seen that the surface of PPTA fiber was smoother after cleaning, and there was no obvious adhesion remaining. Therefore, the latter was more conducive to increasing the bonding force between the fiber and the resin matrix.

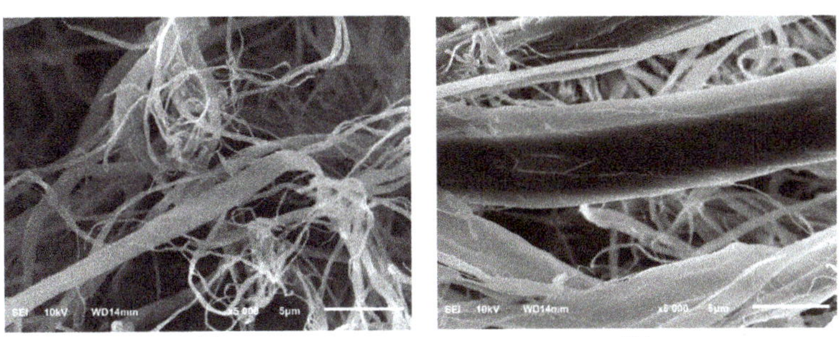

(a) (b)

Figure 3. SEM micrographs of aramid fiber: (**a**) untreated aramid fiber and (**b**) aramid fiber after cleaning with acetone.

In order to further improve the bonding performance of PPTA and PTFE, acetic anhydride was used to modify the surface of PPTA to increase the contact area between PPTA and PTFE matrix [38]. The treatment process was that pour the acetic anhydride solution with a purity specification of analytical reagent into a beaker, and then the PPTA was immersed in the solution. Place the beaker in a water bath and heat to 70 °C and keep it for 1 h. Then, the PPTA was cleaned with ultrapure water and dried for use. The SEM morphology of PPTA fiber surface after acetic anhydride surface treatment is shown in Figure 4.

Figure 4. SEM image of aramid fiber after surface treatment with acetic anhydride for 1 h.

It can be seen from Figure 4 that there were obvious etching marks on the surface of PPTA after modification with acetic anhydride. The surface roughness and the specific surface area of PPTA after etching increased. Therefore, during the molding process of the composite material, PTFE was filled into the cavities on the surface of PPTA fiber, which caused uneven and interlaced interfaces between the fiber and the matrix material after cooling. This not only effectively increased the mechanical chimerism with the PTFE matrix, but also improved the bonding strength between the interfaces.

3.2.2. Preparation of Aramid Fiber and Nano Al_2O_3 Modified PTFE Material

The blending method adopts the mixing method combining dry and wet methods. The detailed steps are as follows. First of all, the pretreated powder and reinforcer were weighed according to the mass percentage, and a certain amount of ethanol solution was poured into a high-speed mixer for mixing. Then, the sieved PTFE powder was slowly poured into the stirring ethanol solution, and then placed in a drying oven at 80 °C for drying and cold-pressed into preformed samples. The preformed sample was put into a box-type electric furnace for sintering to form a composite material sample. Finally, the sample was cut into test specimens with a punching knife for testing. Test samples of composite materials are shown in Figure 5.

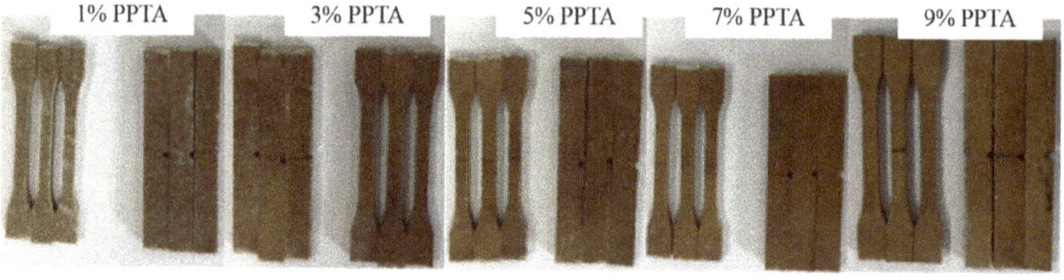

Figure 5. Test sample of PPTA and nano Al_2O_3 filled modified PTFE composite material.

4. Experimental Results and Discussion

4.1. The Structure and Properties of Carbon Material Modified PTFE Composite

4.1.1. Hardness

The impact of different types and contents of carbon materials on the hardness of filled with modified PTFE is shown in Figure 6. During the experiment, we tested each set of the data five times, and calculated the average and standard deviation. It can be seen from Figure 5 that the hardness of PTFE decreased slightly after adding different carbon materials. Among them, the hardness of PTFE composites filled with MWCNT decreased most significantly. When the filling amount of MWCNT was 1%, the hardness of the composite material decreased from 62 to 52, which was 15.3%. With the increase of the content of MWCNT, the hardness rebounds. When the filling amount of 200 nm CF was 1%, the hardness of the composite material dropped from 62 to 58.2, with a decrease of 4.8%. When the filling amount of 7 μm CF was 0.5%, the hardness of the composite material dropped from 62 to 59.6, with a decrease of 5.9%.

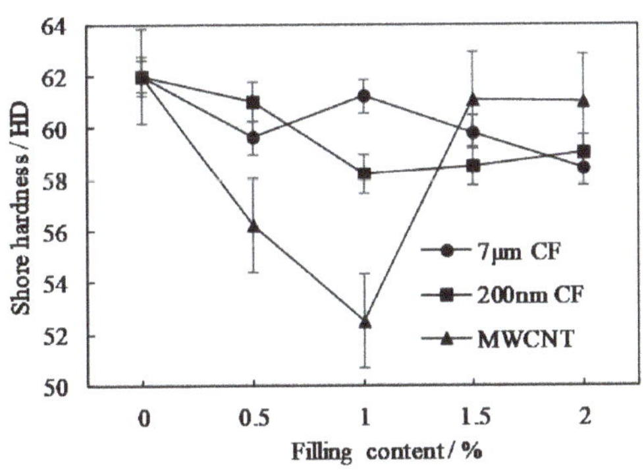

Figure 6. Hardness of micro/nano CF and multiwalled carbon nanotube (MWCNT) filled modified PTFE composites.

4.1.2. Impact Performance

The impact toughness of PTFE composites filled with micro/nano CF and MWCNT was tested. The impact performance of PTFE modified by different carbon materials is shown in Figure 7. It can be seen from the figure that the addition of CF and MWCNT reduced the impact strength of the composite material. Among them, when the filling amount of MWCNT was 1%, the impact toughness of the composite material decreased from 17.89 to 8.73 kJ/m^2, with a decrease rate of 51%. Subsequently, with the increase of the filling amount, the impact toughness tended to stabilize. When the 200 nm CF filling amount was 0.5%, the impact toughness of the composite material decreased from 17.89 to 9.5 kJ/m^2, with a decrease rate of 47%. Subsequently, as the filling amount increased, the impact toughness tended to stabilize. When the filling amount of 7 μm CF was 1%, the impact toughness of the composite material decreased from 17.89 to 8.61 kJ/m^2, with a decrease rate of 55%. Subsequently, as the filling amount increased, the impact toughness slowly rose. The reason is that the mechanical chimerism between the matrix material and the modified material lacked sufficient flexibility, which is prone to brittle fracture when bearing an impact. In addition, the bonding strength between CF and MWCNT and the matrix material was poor, and the energy of an impact fracture cannot be effectively transferred between the composite materials.

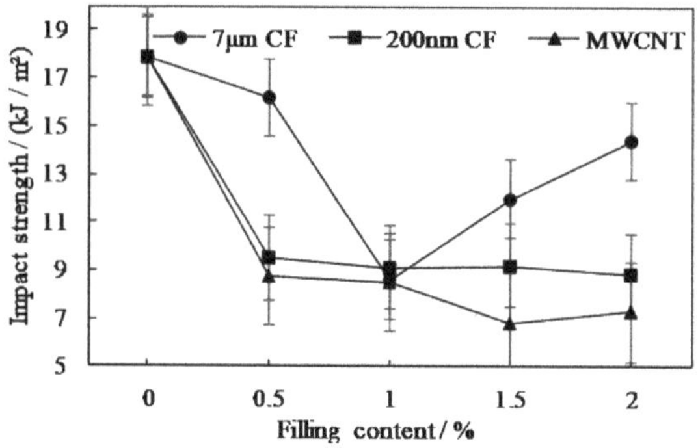

Figure 7. Impact properties of micro/nano CF and MWCNT modified PTFE.

4.1.3. Tensile Performance

The tensile strength, tensile elongation at break and elastic modulus of CF and MWCNT modified PTFE composites with different contents are shown in Figures 8–10.

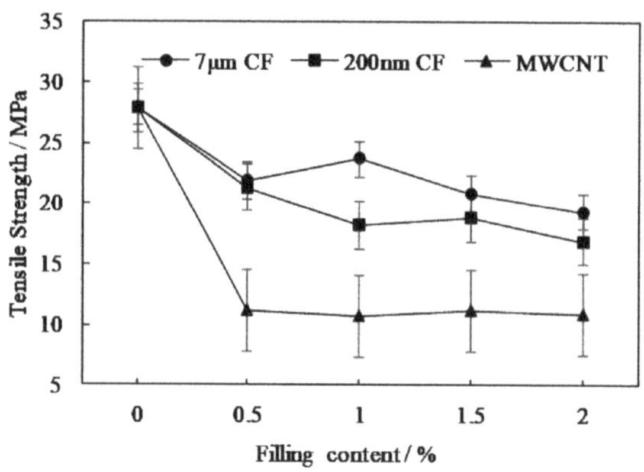

Figure 8. Tensile strength of micro/nano CF and MWCNT modified PTFE composites.

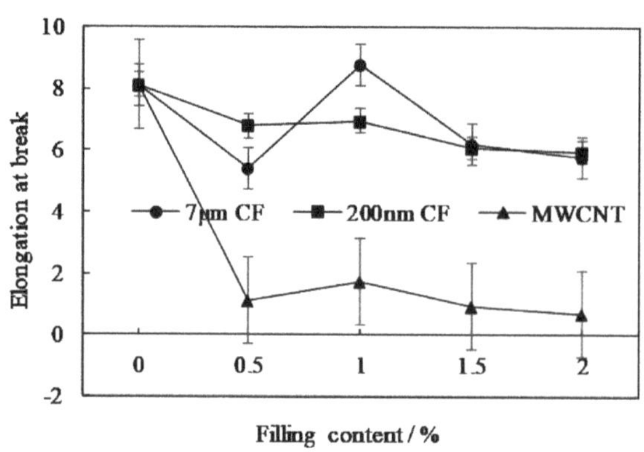

Figure 9. Tensile elongation at break of micro/nano CF and MWCNT modified PTFE composites.

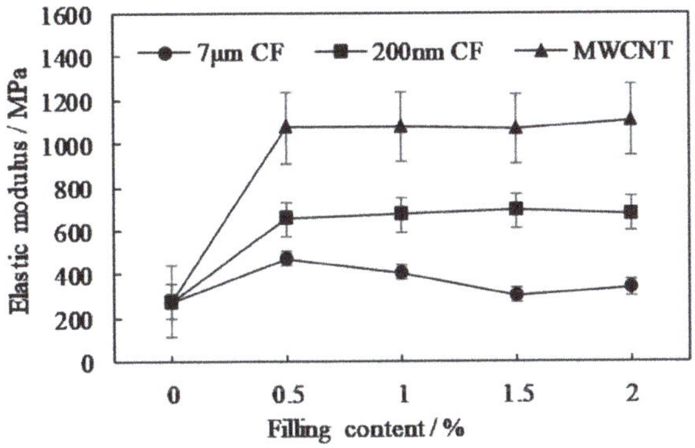

Figure 10. Elastic modulus of micro/nano CF and MWCNT modified PTFE composites.

It can be seen from Figures 8 and 9 that the tensile strength and elongation at break of PTFE were significantly reduced after adding CF and MWCNT. Among them, when the addition amount of 7 μm CF was 0.5%, the tensile strength of the PTFE composite material decreased from 27.9 to 21.87 MPa, which was about 20%, and the tensile elongation at break decreased from 810% to 541%. With the increase of CF to 1%, the tensile strength of the composite material increased to 23.64 MPa, and the tensile elongation at break also rose to the original level. Then, as the CF increased, the tensile strength slowly decreased. Compared with CF, when the addition amount of MWCNT was 0.5%, the tensile strength and tensile elongation at break of PTFE composites dropped significantly to 11.19 MPa and 111%, respectively, with a decrease of about 60% and 87%.

It can be seen from Figure 10 that the elastic modulus of the composite can be effectively improved by adding CF and MWCNT. Among them, the addition of 7 μm and 200 nm CF increased the elastic modulus of PTFE by about 70% to 471.95 MPa and 137% to 655 MPa from the original 276.8 MPa. Then, as the filler increased, the elastic modulus of the composite material tended to stabilize. In contrast, MWCNT had a greater impact on the elastic modulus of PTFE composites. The addition of 1% MWCNT increased the elastic modulus of PTFE from 276.8 to 1075 MPa, with the increase of about 388%. Then, as the filler content increased, the change in the elastic modulus was smaller

By comparing the tensile properties of 200 nm CF filled composites and 7 μm CF filled composites, it can be seen that the elastic modulus of the former had doubled that of the latter, but there was no significant difference between the tensile properties and the tensile fracture productivity. Therefore, it could be concluded that nano-level CF could effectively improve the stiffness of PTFE. When the addition amount of nano CF reached 1.5%, the elastic modulus of the composite material began to decrease, indicating that the excessive CF had a negative effect on the elastic modulus of PTFE. The reason is that with the increase of CF content, CF agglomerates become an impurity point of the composite, which can not further improve its mechanical properties. Compared with the composite filled with CF and MWCNT, it can be seen that MWCNT can significantly enhance the elastic modulus of the composite, but its tensile properties and tensile elongation at break were significantly reduced. This may be because compared with CF, MWCNT had a smaller diameter and more uniform dispersion and larger contact area in PTFE matrix materials. These conditions make it easier to restrict the slippage of PTFE macromolecules, which can effectively increase the elastic modulus of the composite material. However, because the C–C bond in PTFE is wrapped by F atoms, and the F atoms are relatively stable, CF and MWCNT cannot generate a strong chemical bond link with the PTFE matrix, but only through simple physical bonding. During the stretching process, the stress on the PTFE matrix material cannot be transferred to the filler material, which leads to the decrease of the tensile strength of the composite.

In order to further study the influence of different carbon materials and different scale carbon materials on the properties of PTFE composites and the strengthening mechanism, the tensile section of the sample was observed by a scanning electron microscope. In this study, gold was sprayed on the section of the tensile specimen and the surface of the friction and wear specimen, and then the impact section and the wear morphology surface were observed under high magnification using SEM. The SEM of the tensile fracture surface is shown in Figure 11.

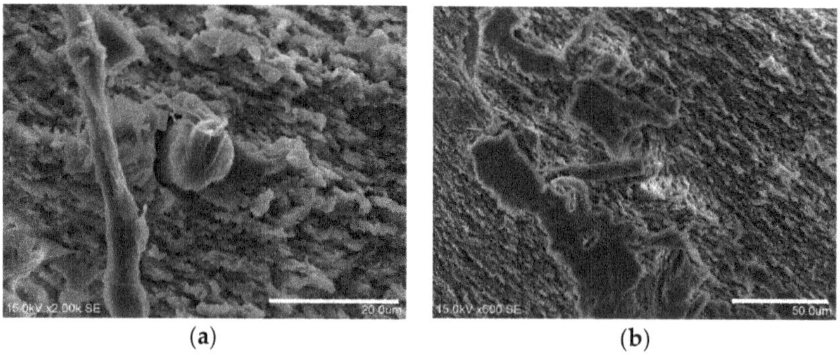

Figure 11. SEM image of the 1% 7 μm CF-filled PTFE tensile section: (**a**) the magnification 2000×; (**b**) the magnification 600×.

It can be seen from Figure 11 that there was CF perpendicular/parallel to the tensile fracture plane. Furthermore, there were certain gaps between the CF and the matrix. These voids easily caused stress concentration, which induced and propagated cracks. Then, the cracks propagated along with the interface, causing the interface to be damaged, and the CF was pulled out of the matrix material, and finally the tensile properties of the composite material were reduced. The fracture mode of the composite material was the fiber/matrix debonding fiber pull-out. Under the action of external force, the interface bonding performance determines the stress transfer effect between the fiber and the matrix, which in turn affected the failure mechanism of the composite material, and ultimately affected the macroscopic mechanical properties of the composite material.

4.1.4. Friction and Wear Performance

The study tested the friction and wear of micro/nano CF and MWCNT filled modified PTFE composites. The friction coefficient and wear rate of the composite material are shown in Figures 12 and 13.

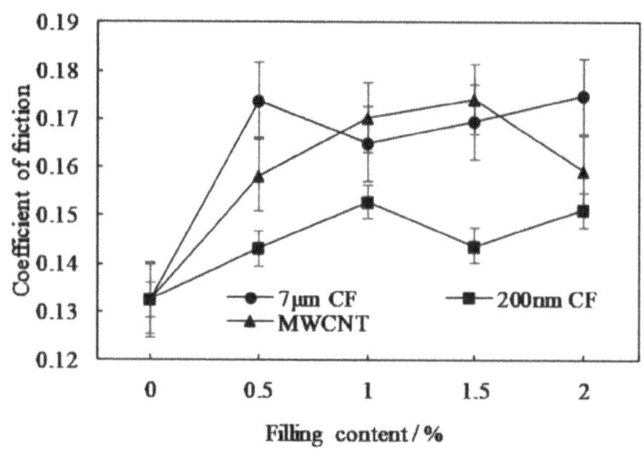

Figure 12. Friction coefficient of micro/nano CF and MWCNT filled modified PTFE composites.

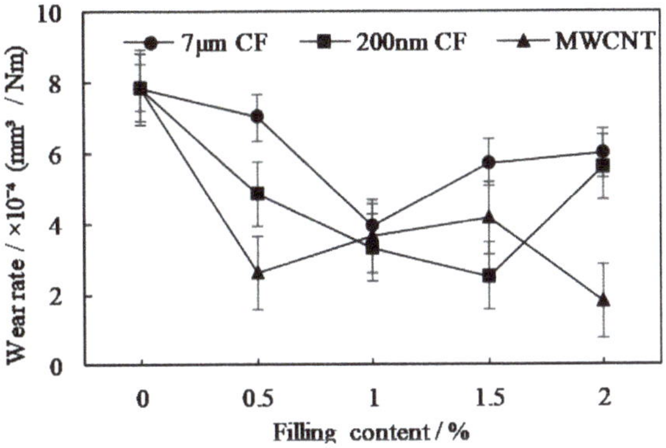

Figure 13. Wear rate of micro/nano CF and MWCNT filled modified PTFE composites.

It can be seen from Figure 12 that the addition of different carbon materials slightly increased the friction coefficient of the composite material. It can be seen from Figure 13 that the strength and load-bearing capacity of the composite material were improved, indicating that the ability of the composite material to resist plastic deformation was significantly improved, and the ploughing effect of the grinding parts on the surface of the composite material was reduced. The addition of CF and MWCNT has greatly improved the wear resistance of composite materials. Under the experimental conditions, as the content of different types of carbon materials increased, the wear of composite materials decreased significantly. Among them, when 7 μm CF was used as the modified filler and the mass fraction was 1.5%, the wear rate of the composite material dropped from 7.8463×10^{-4} to 2.5124×10^{-4} mm^3/Nm, a decrease of 58%. Subsequently, the wear rate of the composite material increased slightly. When 200 nm CF was used as a modified filler and the mass fraction was 1%, the wear rate of the composite material decreased from 7.8463×10^{-4} to 3.9038×10^{-4} mm^3/Nm, a decrease of 50.3%. Then the wear rate gradually increased. When MWCNT was used as a modified filling and the mass fraction was only 0.5%, the wear rate of the composite material decreased from 7.8463×10^{-4} to 2.5897×10^{-4} mm^3/Nm, a decrease of 57%. Subsequently, the wear rate increased slowly, but the wear rate remained below 50% of the unmodified PTFE.

In order to further study the mechanism of filler material on the wear performance of composite materials, the wear surface was tested by scanning electron microscopy. The surface morphology of different carbon materials after wear is observed, as shown in Figure 14.

According to Figure 14, there was no obvious crack on the surface of the composite material. However, there was adhesive wear of the PTFE matrix. In addition, no obvious dents left by carbon CF exposure or peeling of carbon fiber were seen on the PTFE surface. Due to its good self-lubricity, PTFE is easy to form a thin crystalline layer. The crystalline layer and the amorphous layer were alternately arranged in a strip structure. In addition, because of the low friction coefficient, the amorphous partial layer was easy to slide. Since PTFE has a strong intramolecular structure and weak intermolecular binding force, macromolecules are prone to slippage under external force. The essence of PTFE wear is that the macromolecules slip and break under the action of external force, so that the material is pulled out of the crystallization zone and transferred to the surface of the mating part in a sheet shape, resulting in adhesive wear. The addition of CF in PTFE can improve its compressive strength. When the composite is in the process of reciprocating friction and wear, CF mainly plays the role in bearing and limiting the PTFE macromolecular sliding, so it effectively improves the wear resistance of the composite.

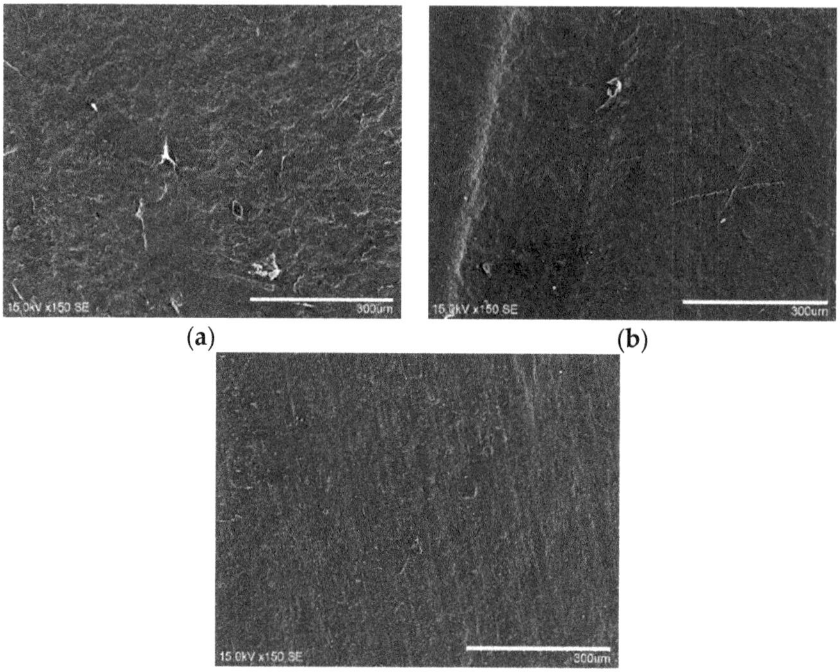

Figure 14. SEM diagram of PTFE wear surfaces filled with different carbon materials: (**a**) 1% 7 μm CF; (**b**) 1% 200 nm CF and (**c**) 1% MWCNT.

4.2. Structure and Properties of the Aramid Modified PTFE Composite

4.2.1. Hardness

The hardness test of PPTA and nano Al_2O_3 filled modified PTFE was conducted. The hardness of the composite material is shown in Figure 15. It can be seen from the figure that after PPTA was added, the hardness of PTFE decreased slightly. When the amount of PPTA added was 1%, the shore hardness of the composite material dropped from 62 to 59.2, a decrease of 4.5%. Subsequently, the hardness of the composite material slowly increased. When the content of PPTA reached the peak at 7%, the hardness of the composite was 61.2, which was similar to the hardness of unmodified PTFE. When 3% nano Al_2O_3 particles and PPTA were compounded and modified into PTFE, the hardness of the PTFE composite material increased slightly from 62 to 64.2 degrees, an increase of 4.8%, indicating that ceramic particles could effectively increase the hardness of PTFE. By comparing the two materials, it can be seen that the hardness of the composite material increased first and then decreased. Three percent nano Al_2O_3 particles could increase the hardness of the composite material on the basis of only filling PPTA.

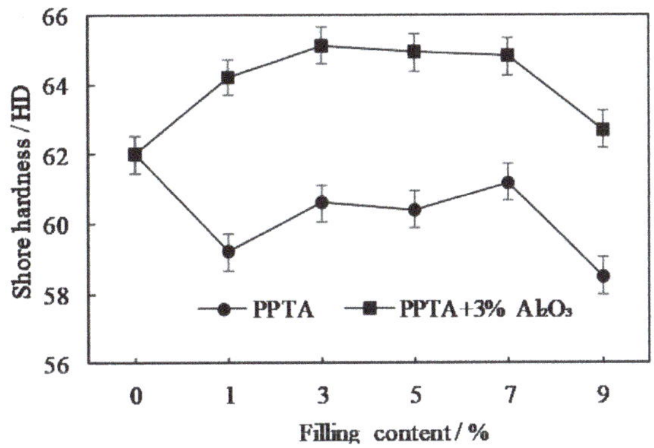

Figure 15. The hardness of PPTA and nano Al_2O_3 particles filled modified PTFE composite material.

4.2.2. Impact Performance

The impact properties of modified PTFE filled with PPTA and nano Al_2O_3 particles were tested. The impact performance of the composite material is shown in Figure 16. It can be seen from the figure that the addition of PPTA significantly reduced the impact toughness of PTFE. When the filling amount of PPTA was 5%, the impact toughness of the composite material decreased from the original 17.89 to 4.46 kJ/m^2, a decrease of 75%, which was due to the poor bonding force between PPTA and Al_2O_3 and the PTFE matrix. In addition, the impact performance of composite materials was mainly affected by fibers, because fibers would significantly affect the failure mode of composite materials during the impact process. The impact strength was sensitive to the notch of the sample, and all impact energy was absorbed in the notch. Therefore, the impact strength had higher requirements for the interface bonding performance of composite materials. However, the impact properties of PPTA composites decreased significantly. The impact strength after adding Al_2O_3 was similar to that of PPTA.

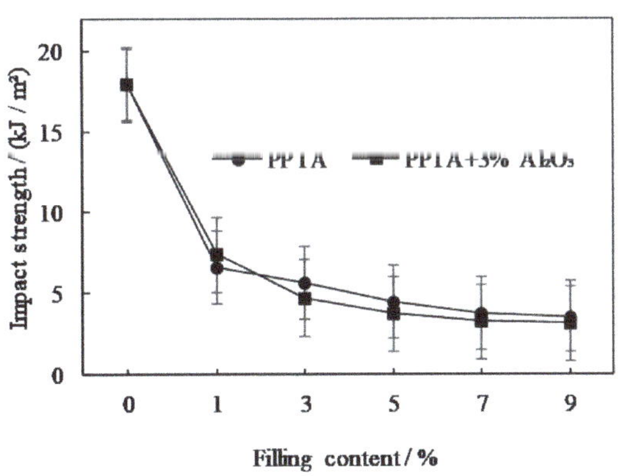

Figure 16. Impact performance of PPTA and nano Al_2O_3 particles filled modified PTFE composites.

4.2.3. Tensile Performance

Tensile properties of PPTA and nano Al_2O_3 particles filled modified PTFE were tested. The tensile strength, tensile elongation at break and elastic modulus of PPTA modified PTFE are shown in Figures 17–19.

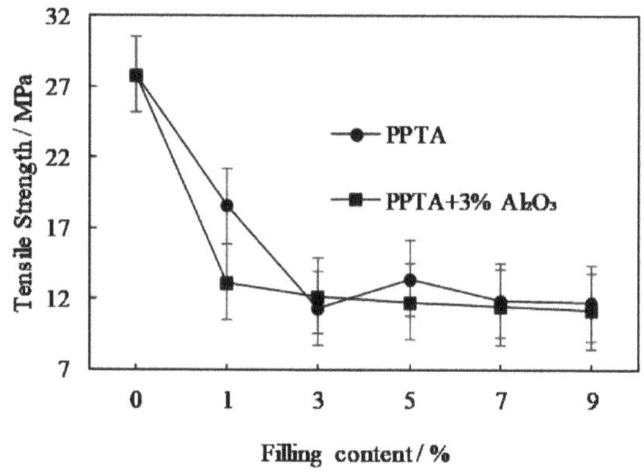

Figure 17. Tensile strength of PPTA and nano Al_2O_3 filled modified PTFE composite.

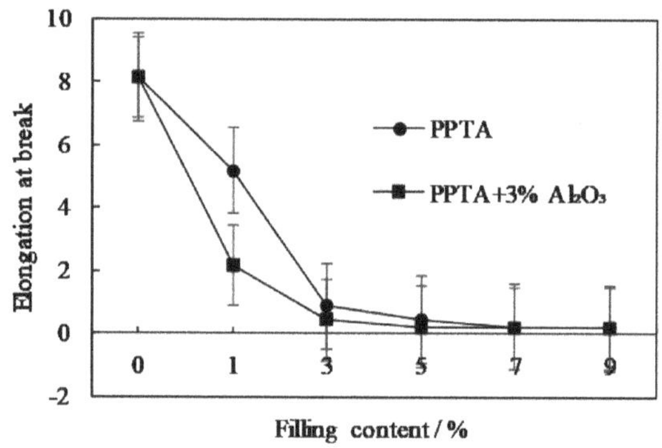

Figure 18. Tensile elongation at break of PPTA and nano Al_2O3 filled modified PTFE composites.

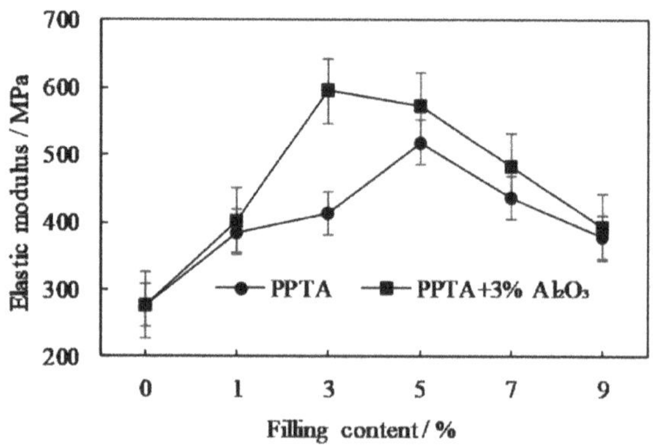

Figure 19. Elastic modulus of PPTA and nano Al_2O_3 filled modified PTFE composite.

It can be seen from Figures 17 and 18 that the tensile strength and elongation at break of PTFE were significantly reduced by adding PPTA. When the PPTA content was 1%, the tensile strength of the composite material dropped from the original 27.9 to 18.6 MPa, with a decrease rate of 33%. At this time, the tensile elongation at break decreased from the original 810–520%, which was only 64% of that of unmodified PTFE at break. When 3% nano Al_2O_3 particles and PPTA were compounded and modified into PTFE, the composite tensile strength further dropped to 13.2 MPa, which was only 50%

of the tensile strength of unmodified PTFE, while the tensile elongation at break decreased to 215%, which was only 26% of unmodified PTFE tensile elongation at break. With the increase of the PPTA content, the tensile strength of the composite material did not change significantly, while the tensile elongation at break further decreased. When the PPTA content was 9%, the tensile elongation at break of the composite material was only 16%. It can be seen from Figure 19 that PPTA and nano Al_2O_3 particles rapidly increased the elastic modulus of PTFE. When PPTA was added at 5%, the elastic modulus reached the highest point of 519 MPa, which was 1.8 times that of unmodified PTFE. With the further increase of PPTA content, the elastic modulus of the composite material gradually decreased. When 3% nano Al_2O_3 particles were compounded and modified with PPTA, the elastic modulus of the composite material was further improved on the basis of only PPTA. Among them, when the addition amount of nano Al_2O_3 particles and PPTA was 3%, the elastic modulus reached the highest point 595 MPa.

With the increase of PPTA, PPTA and PTFE in the composite material were entangled with each other, resulting in a rapid increase in the elastic modulus of the composite material. When the amount of PPTA added was 5%, the elastic modulus began to decrease gradually. This is because when the content of PPTA was high, the PTFE matrix material could not completely wrap the PPTA fibers, resulting in contact between some PPTA fibers and PPTA fibers. However, there was only an electrostatic force between PPTA and PPTA as the binding force, which led to the decrease of composite properties. It can be seen from the tensile strength that the addition of PPTA significantly reduced the tensile properties of PTFE, because the C–C bond on the PTFE surface was wrapped by F atoms, and the F atoms were relatively stable. Thus, the PPTA fiber and the PTFE matrix interface could not form a strong bonding force, only the interaction of physical meshing and electrostatic force was generated. Moreover, the bonding force was relatively weak, resulting in the gradual decline of the tensile properties of the material.

In order to further study the influence of PPTA on the properties of PTFE and the strengthening mechanism, the tensile and impact sections of the sample were observed by SEM. The SEM images of the tensile fracture surface and the impact fracture surface are shown in Figure 20. It can be seen from the figure that the PPTA fiber and PTFE matrix material were entangled with each other. The impact fracture surface presents ductile fracture, and the interface bond lacked flexibility. It can be seen from the fracture surface that the PPTA thread that was obviously drawn from the other fracture surface protrudes from the fracture surface, indicating that the bonding performance between PPTA and the PTFE matrix was poor. When impact fracture or tensile fracture occurred, there was a clear boundary layer between PPTA and PTFE. Due to the small interaction between PTFE matrix and PPTA, PPTA will pull out from PTFE and eventually lead to the fracture of the composite.

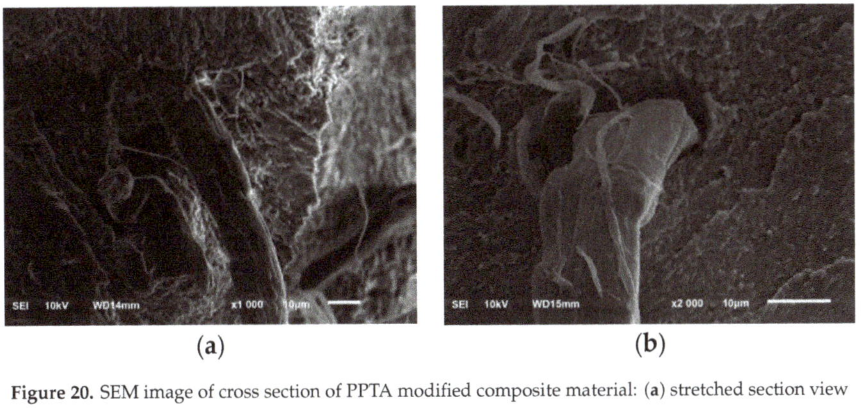

Figure 20. SEM image of cross section of PPTA modified composite material: (**a**) stretched section view and (**b**) impact section.

4.2.4. Friction and Wear Performance

In the study, the friction and wear properties of PPTA and nano Al_2O_3 filled modified PTFE were tested. The friction coefficient and wear rate of the composite material are shown in Figures 21 and 22. It can be seen from Figure 21 that the addition of PPTA and nano-Al_2O_3 particles improved the friction coefficient of the composite material. It can be seen from Figure 22 that the wear rate of the composite material was significantly improved. When the PPTA filling amount was 1%, the wear rate of the composite material dropped from the original 7.85×10^{-4} to 5.18×10^{-4} mm^3/Nm, a decrease of 35%. It can be found that the ability of the composite material to resist plastic deformation was significantly improved, and the ploughing effect of the grinding parts on the surface of the composite material was reduced. The addition of PPTA could effectively reduce material loss and improve friction and wear performance. With the increase in the mass fraction of PPTA, the friction coefficient of the composite material increased slightly by 6–12%, mainly because the composite material formed a thicker transfer film during the adhesive wear process. With the addition of PPTA, the friction torque increased and the friction coefficient also increased. In addition, with the increase of PPTA content, the wear rate of the composite material decreased first and then rose. Among them, when the content of PPTA was 5%, the wear rate of the composite material was the lowest at 2.47×10^{-4} mm^3/Nm, with a decrease rate of 68%.

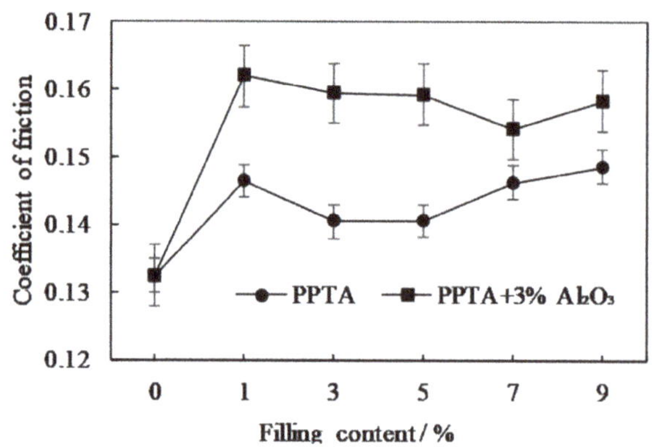

Figure 21. Wear coefficient of modified PTFE composite filled with PPTA and nano Al_2O_3 particles.

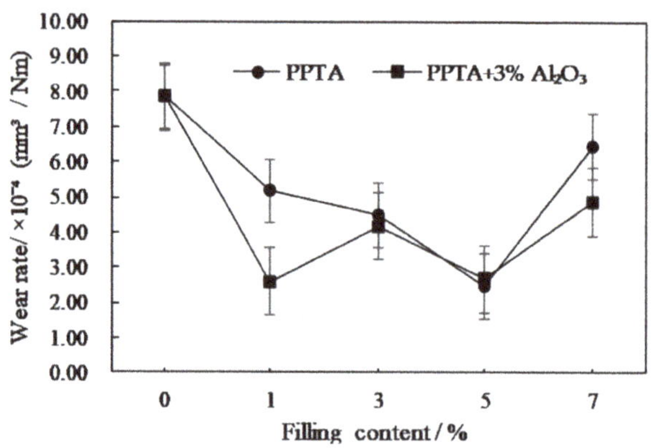

Figure 22. Wear rate of PPTA and nano Al_2O_3 particles filled modified PTFE composite.

When the content of PPTA exceeded a certain amount, the wear rate of the composite material presented an instantaneous surge. This is mainly because the composite material mainly adhered to each other through the PTFE matrix and transferred the load to the fibers. Due to the large specific surface area of PPTA, when PPTA increases to a certain content, the PTFE matrix will not be able to bond with PPTA well. However, the interaction between PPTA is mainly through an electrostatic force, so the fibers are easy to be peeled off and pulled out, which makes the overall performance of the composite material drop sharply.

In order to further study the wear mechanism of the composites, the wear surface was tested by SEM and the surface morphology after wear was observed. The wear morphology of the composite material under dry friction conditions is shown in Figure 23.

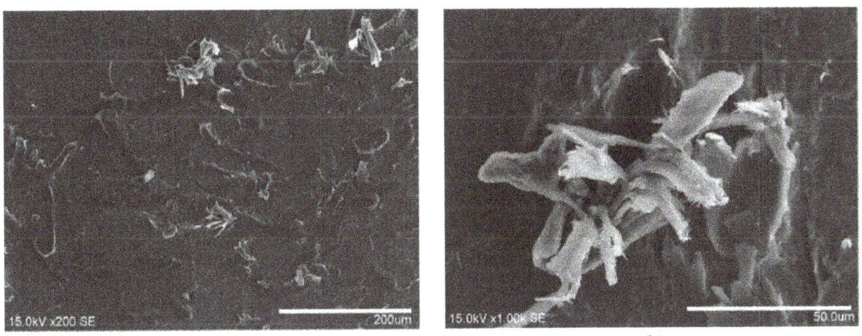

Figure 23. SEM image of the wear surface of 5% PPTA filled PTFE: (**a**) the magnification 200×; (**b**) the magnification 1000×.

It can be seen from Figure 23 that there were obvious PPTA fibers distributed, and a little PPTA pulled out on the wear surface. The wear surface of the composite material was rough. PTFE exhibited large flakes, without obvious adhesive wear. The PPTA fiber was exposed on the wear surface of the composite material and had poor bonding performance with the matrix material. When PPTA was added into PTFE, PPTA mainly played the role of bearing and restraining the slippage of PTFE macromolecules, so as to effectively improve the wear resistance of the composite material.

5. Conclusions

(1) Through the ultrasonic dispersion of CF and MWCNT and various processes (crushing, stirring volatilization, crushing and sieving) of PTFE materials, the effective dispersion of CF and MWCNT in the composite material and the stability of the composite material performance obtained an effective guarantee.

(2) Through testing the hardness and impact performance of CF and MWCNT filled modified PTFE composites, it was found that various carbon materials would cause the hardness and impact toughness of PTFE composites to decrease to varying degrees. Among them, the hardness of MWCNT-filled PTFE composites decreased most obviously. When the content of MWCNT was 1%, the impact hardness of the composite material dropped from 62 to 52, with a decrease of 15.3%.

(3) By testing the tensile properties of CF and MWCNT filled modified PTFE composites, it was found that the addition of various carbon materials would significantly reduce the tensile strength and elongation at break of PTFE, but greatly enhanced the elastic modulus. Among them, the addition of 7 μm CF reduced the tensile strength of the PTFE composite material from 27.9 MPa by about 20%, and the tensile elongation at break decreased by about 30%. The addition of MWCNT significantly reduced the tensile strength and tensile elongation at break of PTFE composites, by 60% and 87% respectively. Among them, the addition of 7 μm and 200 nm CF increased the elastic modulus of PTFE by about 70% to 471.95 MPa and 137% to 655 MPa from the original 276.8 MPa. MWCNT had a greater impact on the elastic modulus of PTFE composites.

The addition of 1% MWCNT increased the elastic modulus of PTFE from 276.8 to 1075 MPa, with an increase of 388%.

(4) Through the friction and wear test of 200 nm and 7 μm CF and MWCNT filled PTFE composite materials, it was found that when 7 μm CF was used as the modified filler and the mass fraction was 1.5%, the composite wear rate decreased from 7.8463×10^{-4} to 2.5124×10^{-4} mm^3/Nm, with a decrease rate of 58%. Then the wear rate of the composite material increased slightly. When 200 nm CF was used as the modified filler and the mass fraction was 1%, the wear rate of the composite material decreased from 7.8463×10^{-4} to 3.9038×10^{-4} mm^3/Nm, with a decrease rate of 50.3%. Then the wear rate gradually increased. When MWCNT was used as a modified filling and the mass fraction was only 0.5%, the wear rate of the composite material decreased from 7.8463×10^{-4} to 2.5897×10^{-4} mm^3/Nm, with a decrease rate of 57%.

(5) Through the hardness and impact toughness test of PPTA and nano Al_2O_3 particle composite modified PTFE, it was found that the increase of PPTA and nano Al_2O_3 particles reduced the hardness and impact toughness of PTFE. When the PPTA content was 1%, the shore hardness of the composite material dropped from 62 to 59.2, with a decrease rate of 4.5%. When the content of PPTA was 5%, the impact hardness of the composite material dropped from the original 17.89 to 4.46 kJ/m^2, with a decrease of 75%.

(6) Through the tensile performance test of PPTA and nano Al_2O_3 particles composite modified PTFE, it was found that the increase of PPTA and nano Al_2O_3 particles could effectively increase the elastic modulus of PTFE. Among them, PPTA composite modification had a more obvious increase in the elastic modulus of PTFE. As the content of PPTA increased, the elastic modulus of the composite material first rose and then fell. When PPTA was added at 5%, the elastic modulus reached the highest value of 519 MPa. With the increase of filler content, the tensile strength and tensile elongation of PTFE composite material modified by PPTA fiber and nano-Al_2O_3 decreased.

(7) By testing the friction and wear properties of PPTA and nano-Al_2O_3 composite modified PTFE, it was found that the addition of PPTA and nano-Al_2O_3 particles could significantly increase the wear rate of the composite. When the PPTA filling amount was 1%, the wear rate of the composite material dropped from the original 7.85×10^{-4} to 5.18×10^{-4} mm^3/Nm, with a decrease of 35%.

Author Contributions: Conceptualization, F.Z. and Y.Z.; validation, J.Z. and Y.J.; formal analysis, F.Z. and X.W.; resources, F.Z.; writing—original draft, J.Z.; writing—review and editing, F.Z.; visualization, J.Z.; supervision, F.Z. All authors have read and agreed to the published version of the manuscript.

Funding: This research was supported by A Priority Academic Program Development of Jiangsu Higher Education Institutions (PAPD); Engineering Research Center for Harmless Treatment and Resource Utilization of Aluminum Ash and Slag Solid Waste in Jiangsu Province; Nantong Applied Research Project, grant number JC2018115 and JC2019129; Ministry of Education's Cooperative Education Project, grant number 201901189019.

Acknowledgments: Sincere thanks go to the anonymous reviewers for their valuable comments and good suggestions to improve this article.

Conflicts of Interest: The authors declare no conflict of interest.

References

1. Sato, Y.; Murahara, M. Protein adsorption on PTFE surface modified by ArF excimer laser treatment. *J. Adhes. Sci. Technol.* **2004**, *18*, 1545–1555. [CrossRef]
2. Liu, Y.; Xu, N.; Wang, Y.; Yao, Y.; Xiao, H.; Jia, J.; Lv, H.; Zhang, D. Preparation and tribological properties of hybrid PTFE/Kevlar fabric self-lubricating composites. *Surf. Coat. Technol.* **2019**, *361*, 196–205. [CrossRef]
3. Ashrafi, F.; Lavasani, M.R.B. Improvement of the mechanical and thermal properties of polyester nonwoven fabrics by PTFE coating. *Turk. J. Chem.* **2019**, *43*, 760–765. [CrossRef]
4. Villalobos, G.J.; Dow, N.; Milne, N.; Zhang, J.; Naidoo, L.; Gray, S.; Duke, M. Membrane distillation trial on textile wastewater containing surfactants using hydrophobic and hydrophilic-coated polytetrafluoroethylene (PTFE) membranes. *Membranes* **2018**, *8*, 31. [CrossRef]

5. He, G.; Zeng, F.; Lei, K.; Xia, S.; Deng, L.; Zhang, C.; Liu, D. Handle operation lug type ptfe lined butterfly valve (TD71X). *Eur. J. Hosp. Pharm.* **2017**, *24*, 162. [CrossRef]
6. Haque, R.I.; Briand, D. Triboelectric freestanding flapping film generator for energy harvesting from gas flow in pipes. *Smart Mater. Struct.* **2019**, *28*, 085002. [CrossRef]
7. Feng, S.; Zhong, Z.; Wang, Y.; Xing, W.; Drioli, E. Progress and perspectives in PTFE membrane: Preparation, modification, and applications. *J. Membr. Sci.* **2018**, *549*, 332–349. [CrossRef]
8. Yanhai, C.; Lu, R.; Xianliang, M.; Jinyong, Y.; Xianhua, T. The effect of PTFE addition on mechanical and anti-corrosion properties of coating of heat exchangers. *Mater. Res. Express* **2019**, *6*, 085207. [CrossRef]
9. Wang, J.; Wu, D.; Li, X.; Zhang, M.; Zhou, W. Poly (vinylidene fluoride) reinforced by carbon fibers: Structural parameters of fibers and fiber-polymer adhesion. *Appl. Surf. Sci.* **2012**, *258*, 9570–9578. [CrossRef]
10. Luo, F.; Tang, B.; Fang, Z.; Yuan, Y.; Li, H.; Zhang, S. Polytetrafluoroethylene based, F8261 modified realization of $Li_2SnMg_{0.5}O_{3.5}$ filled composites. *Appl. Surf. Sci.* **2020**, *503*, 144088. [CrossRef]
11. Shi, Y.; Feng, X.; Wang, H.; Lu, X. The effect of surface modification on the friction and wear behavior of carbon nanofiber-filled PTFE composites. *Wear* **2008**, *264*, 934–939. [CrossRef]
12. Gürgen, S.; Çelik, O.N.; Kuşhan, M.C. Tribological behavior of UHMWPE matrix composites reinforced with PTFE particles and aramid fibers. *Compos. B Eng.* **2019**, *173*, 106949. [CrossRef]
13. Yao, S.S.; Jin, F.L.; Rhee, K.Y.; Hui, D.; Park, S.J. Recent advances in carbon-fiber-reinforced thermoplastic composites: A review. *Compos. B. Eng.* **2018**, *142*, 241–250. [CrossRef]
14. Ni, H.; Zhang, J.; Lv, S.; Wang, X.; Pei, Y.; Li, F. Coating process parameters and structural properties of the tubular electrodes of fuel cells based on a self-made coating device. *Coatings* **2020**, *10*, 830. [CrossRef]
15. Subramanian, R.; Shanmugam, K.; Marappan, S. Fabrication of robust superhydrophobic coatings using PTFE-MWCNT nanocomposite: Supercritical fluid processing. *Surf. Interface Anal.* **2018**, *50*, 464–470. [CrossRef]
16. Xu, F.; Fan, W.; Zhang, Y.; Gao, Y.; Jia, Z.; Qiu, Y.; Hui, D. Modification of tensile, wear and interfacial properties of Kevlar fibers under cryogenic treatment. *Compos. B Eng.* **2017**, *116*, 398–405. [CrossRef]
17. Liu, K.; Yang, W.; Peng, T.; Wang, F. Progress in Surface Modification of Aramid Fiber (II)—Chemical Modification. *Synth. Fiber China* **2011**, *40*, 26–31. (In Chinese) [CrossRef]
18. Gonzalez-Canche, N.G.; Flores-Johnson, E.A.; Carrillo, J.G. Mechanical characterization of fiber metal laminate based on aramid fiber reinforced polypropylene. *Compos. Struct.* **2017**, *172*, 259–266. [CrossRef]
19. Ni, H.; Zhang, J.; Lv, S.; Gu, T.; Wang, X. Performance optimization of original aluminum ash coating. *Coatings* **2020**, *10*, 831. [CrossRef]
20. Ni, H.; Zhang, J.; Lv, S.; Wang, X.; Zhu, Y.; Gu, T. Preparation and performance optimization of original aluminum ash coating based on plasma spraying. *Coatings* **2019**, *9*, 770. [CrossRef]
21. Shuaishuai, L.; Jiaqiao, Z.; Hongjun, N.; Xingxing, W.; Yu, Z.; Tao, G. Study on preparation of aluminum ash coating based on plasma spray. *Appl. Sci.* **2019**, *9*, 4980.
22. Vasilev, A.P.; Okhlopkova, A.A.; Struchkova, T.S. Investigation of the influence of complex fillers on the properties and structure of polytetrafluoroethylene. *J. Frict. Wear* **2018**, *39*, 427–432. [CrossRef]
23. Ohlopkova, A.A.; Struchkova, T.S.; Vasilev, A.P. Studying the properties and structure of polytetrafluoroethylene filled with Belum modified carbon fibers. *J. Frict. Wear* **2016**, *37*, 529–534. [CrossRef]
24. Bajpai, A.; Saxena, P.; Kunze, K. Tribo-mechanical characterization of carbon fiber-reinforced cyanate ester resins modified with fillers. *Polymers* **2020**, *12*, 1725. [CrossRef]
25. Saxena, P.; Schinzel, M.; Andrich, M. Development of a novel test-setup for identifying the frictional characteristics of carbon fibre reinforced polymer composites at high surface pressure. *IOP Conf. Ser. Mater. Sci. Eng.* **2016**, *149*, 012124. [CrossRef]
26. Luo, F.; Tang, B.; Yuan, Y. Microstructure and microwave dielectric properties of $Na_{1/2}Sm_{1/2}TiO_3$ filled PTFE, an environmental friendly composites. *Appl. Surf. Sci.* **2018**, *436*, 900–906. [CrossRef]
27. Yuan, Y.; Li, Z.; Cao, L. Modification of Si_3N_4 ceramic powders and fabrication of Si_3N_4/PTFE composite substrate with high thermal conductivity. *Ceram. Int.* **2019**, *45*, 16569–16576. [CrossRef]
28. Van Rooyen, L.J.; Bissett, H.; Khoathane, M.C. Gas barrier properties of oxyfluorinated graphene filled polytetrafluoroethylene nanocomposites. *Carbon* **2016**, *109*, 30–39. [CrossRef]
29. GB/T 2411-2008. *Plastics and Ebonite-Determination of Indentation Hardness by Means of a Durometer (Shore Hardness)*; China National Standardization Management Committee: Beijing, China, 2008. (In Chinese)
30. GB/T1040-2006. *Determination of Tensile Properties of Plastics*; China National Standardization Management Committee: Beijing, China, 2006. (In Chinese)

31. GB/T1043-2008. *Determination of Impact Properties of Plastic Simply Supported Beams;* China National Standardization Management Committee: Beijing, China, 2008. (In Chinese)
32. Jiang, B.; Zhu, A.; Zhang, C.; Li, Y. Interface enhancement between polytetrafluoroethylene and glass fibers modified with a titanate coupler. *J. Appl. Polym. Sci.* **2017**, 134. [CrossRef]
33. Lv, S.; Zhang, J.; Ni, H.; Wang, X.; Zhu, Y.; Chen, L. Study on the coupling relationship of low temperature fluidity and oxidation stability of biodiesel. *Appl. Sci.* **2020**, *10*, 1757. [CrossRef]
34. Paul, R.; Bhowmik, S. Tribological Behavior of micro coir filler reinforced polymer composite under dry, wet, and heated contact condition. *J. Nat. Fibers* **2020**, 1–16. [CrossRef]
35. Zhang, J.; Kang, X.; Ni, H.; Ren, F. Surface defect detection of steel strips based on classification priority YOLOv3-dense network. *Ironmak. Steelmak.* **2020**, 1–12. [CrossRef]
36. Ni, H.; Wang, K.; Lv, S.; Wang, X.; Zhuo, L.; Zhang, J. Effects of concentration variations on the performance and microbial community in microbial fuel cell using swine wastewater. *Energies* **2020**, *13*, 2231. [CrossRef]
37. Ni, H.; Wang, K.; Lv, S.; Wang, X.; Zhang, J.; Zhuo, L.; Li, F. Effects of modified anodes on the performance and microbial community of microbial fuel cells using swine wastewater. *Energies* **2020**, *13*, 3980. [CrossRef]
38. Wang, L.; Duan, Y.; Zhang, Y.; Huang, R.; Dong, Y.; Huang, C.; Zhou, B. Surface modification of poly-(p-phenylene terephthalamide) pulp with a silane containing isocyanate group for silicone composites reinforcement. *Wuhan Univ. J. Nat. Sci.* **2016**, *21*, 505–511. [CrossRef]

coatings

Article

Analysis of Surface Properties of Ag and Ti Ion-Treated Medical Textiles by Metal Vapor Vacuum Arc Ion Implantation

Ali Akpek [1,2]

1 Department of Bioengineering, Faculty of Engineering, Gebze Technical University, 41400 Kocaeli, Turkey; aliakpek@gtu.edu.tr
2 Sabanci University Nanotechnology Research and Application Center, Sabanci University, 34956 Istanbul, Turkey

Abstract: The study focuses on the effects of Ag (silver) and Ti (titanium) ions on textiles by MEVVA (metal vapor vacuum arc) ion implantation. In order to comprehend this, the research was executed in three parts. In the first part, the antibacterial efficiencies of Ag and TiO_2 were investigated in detail since the antibacterial capabilities of Ag and TiO_2 are well known. A group of polyester- and cotton-based medical textiles were modified by Ag and TiO_2 ions, with doses ranging from 5×10^{15} to 5×10^{16} ion/cm^2. To determine the adhesion capabilities of the implanted ions on surfaces, after the first round of antibacterial tests, these medical textiles were washed 30 times, and then antibacterial tests were performed for the second time. The results were also compared with nanoparticle-treated medical textiles. In the second part, the corrosion and friction capabilities of Ag and Ti ion-implanted polyester textiles, with a dose of 5×10^{15} ion/cm^2, were investigated. Finally, the UV protection capabilities of Ag and Ti ion-implanted polyester textiles, with a dose of 5×10^{15} ion/cm^2, were investigated. The experiments showed that even after 30 washes, the TiO_2 ion-implanted polyester textile had almost 85% antibacterial efficiency. In addition, Ti ion implantation reduced the friction coefficiency of a polyester textile by almost 50% when compared with an untreated textile. Finally, the Ag-ion-implanted polyester textile provided a UV protection factor of 30, which is classified as very good protection.

Keywords: surface engineering; ion implantation; MEVVA; antibacterial, corrosion, UV protection

Citation: Akpek, A. Analysis of Surface Properties of Ag and Ti Ion-Treated Medical Textiles by Metal Vapor Vacuum Arc Ion Implantation. *Coatings* **2021**, *11*, 102. https://doi.org/10.3390/coatings11010102

Received: 30 November 2020
Accepted: 12 January 2021
Published: 18 January 2021

1. Introduction

As nanotechnology becomes a part of our daily lifestyle, nanotextile products have become one of the most important gains of this new progress. Each and every day, international corporations offer new nanotextile products to the market, and, every day, these products get a bigger market share. Stain-resistant, liquid-repellent, odor-resistant, antimosquito, and antistress textiles have already been made available everywhere in the world. Nowadays, one of the most important topics in the nanotextile area is antibacterial research [1,2].

International corporations and scientists have been working for almost 30 years on antibacterial textiles and developing new products [3,4]. However, none of these products' antibacterial ability has been successful, cheap, unlimited or briefly useful enough for customers. Due to this reason, only a small percentage of the world population use antibacterial technologies for their clothes.

On the other hand, resistant pathogens have created a major problem for health services. As an example, *Staphylococcus aureus* can develop resistance to methicillin and, because of this, each and every year, millions of people die from this deathly hospital infection. To reduce the death rate of this infection, hospital clothes and wound dressings that include silver particles were developed. It is very well known that silver ions and silver-based compounds are highly toxic to microorganisms [5]. However, even these novel products are not satisfactory enough.

There have been several attempts to create antibacterial textiles. Li et al. reported the potential use of alginate for antibacterial textile production [6]. In a similar approach Turakhia et al. developed copper oxide nanoparticles on cotton fabrics [7], Zemjic et al. developed chitosan nanoparticles [8], Fouda et al. developed zinc oxide nanoparticles [9] and Ouadil et al. prepared Ag-loaded graphene nanocomposites for antibacterial medical textile development [10]. All these studies have claimed to have 100% antibacterial ability at the end of the research. However, since nanoparticles are not immune to laundering their antibacterial ability after several washouts is still unknown.

For all these reasons, new methodologies have been investigated to make nanotextiles cheaper and more successful. In addition to that, it is very important for these improved textiles to preserve their antibacterial efficiencies even after several rounds oflaundering. Previous studies have never provided this information. It is also obvious that one particular surface modification technique is one step ahead of all other techniques. This technique is called ion beam implantation technology.

There are three methods commonly used for ion implantation. These methods are mass-analyzed ion implantation (MAII), direct ion implantation (DII), and plasma source ion implantation (PSII). One of the most recent ion implantation techniques is metal vapor vacuum arc (MEVVA), a type of DII ion source implantation. The method used in this study is MEVVA ion beam implantation technology.

Ion implantation is a green, environmental-friendly surface engineering technology. During ion implantation, processed heat on the surface of the material is between 40 and 100 °C. Therefore, no corruption of any material can be observed between this heat range. The main principle of ion beam implantation technology relies on accelerating and bombarding ions to the surface of the targeted material under vacuum and providing them improved surface properties. Collided ions do not affect the bulk structure of the main material. The main mechanical properties of the modified materials are protected. The MEVVA source incorporates a cathode, where cathode materials and ion species can quickly be generated and put into the ion extractor. The source also includes a powerful magnetic field and injection of gas into the ion extractor. Then, the desired ion beams can be generated. The voltage of the ion source extraction can be a maximum of 100 kV. The ion beam is a broad beam of 10 cm after the extraction. The radial beam redirects with magnets and evolves into a broader beam until the ion acceleration column. Finally, the accelerated ions arrive at the end station and are implanted into the surface material. This entire process is explained in Figure 1.

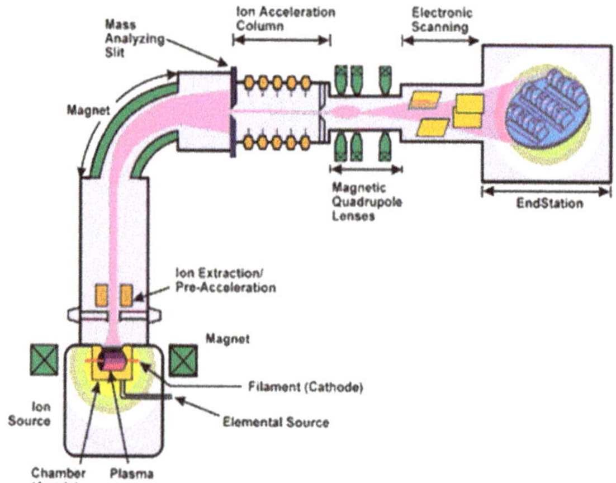

(a)

Figure 1. *Cont.*

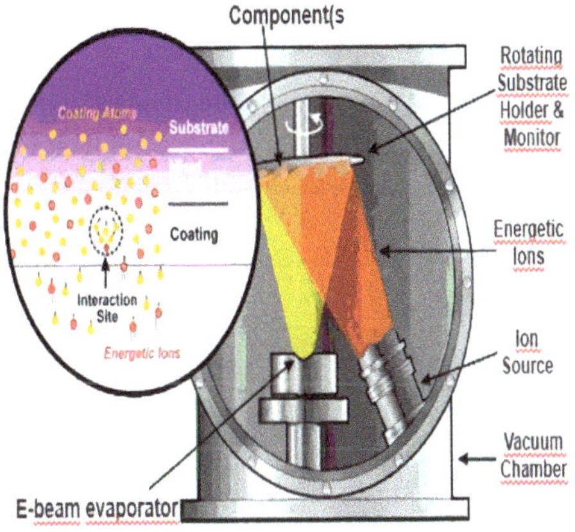

(b)

Figure 1. (**a**) The creation and acceleration of ion beams. (**b**) Ion implantation inside surface materials [11,12].

Ion implantation is a surface modification technique that enables us to modify surfaces with such desired abilities as corrosion resistance, friction resistance, biocompatibility, hydrophilia, hydrophobia, and antibacterial efficiency. These are only some of the abilities that ion implantation technology can provide metal, ceramic, plastic, and polymer material surfaces [13]. Unfortunately, there are only a few studies where MEVVA has been used for textiles [13–15]. The only study where MEVVA was used for creating antibacterial ability was achieved by us with very limited resources. That research pioneered our current research. As an addition to our previous studies, in this research, the antibacterial effects of Ag and TiO_2 are investigated more precisely, SEM analysis is achieved, UPF factor and friction resistance capabilities are also analyzed, and all data are categorized as evidence of the general effects of Ag and TiO_2, especially on medical textiles. Therefore, this research may have a serious impact on the future development of smart textiles that are specifically for medical purposes.

Ag (silver) and TiO_2 (titanium dioxide) are known for their antibacterial abilities [16,17]. Due to these reasons, these ions were selected for implantation. Friction and UV protection capabilities were also analyzed to figure out the other advantages or disadvantages of Ag- and Ti-implanted polyester fabrics. There is enough evidence in the literature to claim that these ions may contribute seriously to the friction and UV protection capabilities of materials [18,19]. Therefore, these tests were also conducted on ion-implanted polyester textiles.

This study is one of the first studies to use MEVVA ion beam implantation technology as an alternative nanotextile technology. To evaluate its potential commercial value, the results were compared with the antibacterial efficiency results of commonly used and commercially available nanoparticles. Additionally, in order to determine the adhesion capabilities of the implanted ions and nanoparticles treated on the surfaces, after the first antibacterial tests, these medical textiles were washed 30 times, and antibacterial tests were performed a second time. The results were compared afterward.

2. Experiments

2.1. Ion-Implanted Fabrics

The cotton- and polyester-based medical textiles (Labor Textile Inc., Izmir, Turkey) were cut into a 10 × 10 cm^2 size, with 0.1 gr/cm^2, and sent to the MEVVA ion beam implantation laboratory. In this article, cotton fabric refers to 100% alpaca cotton fabric and polyester fabric refers to 75% polyester fabric with 25% cotton. These textiles were selected since they are the most commonly used medical textiles in the market [20]. The textiles were fabricated for medical use in the first place.

A vacuum arc ion source (MEVVA) based metal ion implantation system was used on the fabrics to enhance their surface properties. An MEVVA system is shown in Figure 2. This MEVVA source utilizes the principle of vacuum arc discharge between the cathode and the anode to create dense plasma from which an intense beam of metal ions of the cathode material is extracted. This metal ion source is operated in pulse mode. In our MEVVA system, when Ag was used, the ion source was adjusted to pulse with 50 Hz frequency for 250 μs time duration (ζ) and an ion beam current of 2 mA/cm^2. The device worked under 4×10^{-6} Torr pressure. When Ti or TiO_2 was used to obtain antibacterial ability, the ion source was adjusted to pulse with 50 Hz frequency for 250 μs time duration (ζ) and an ion beam current of 1 mA/cm^2. The device worked under 5×10^{-6} Torr pressure. Titanium ions and oxygen ions were accelerated and implanted individually, and, for each of them, previous values were used in this study. When oxygen ions are implanted after titanium ions, they spontaneously merge with each other and form TiO_2 (titanium dioxide).

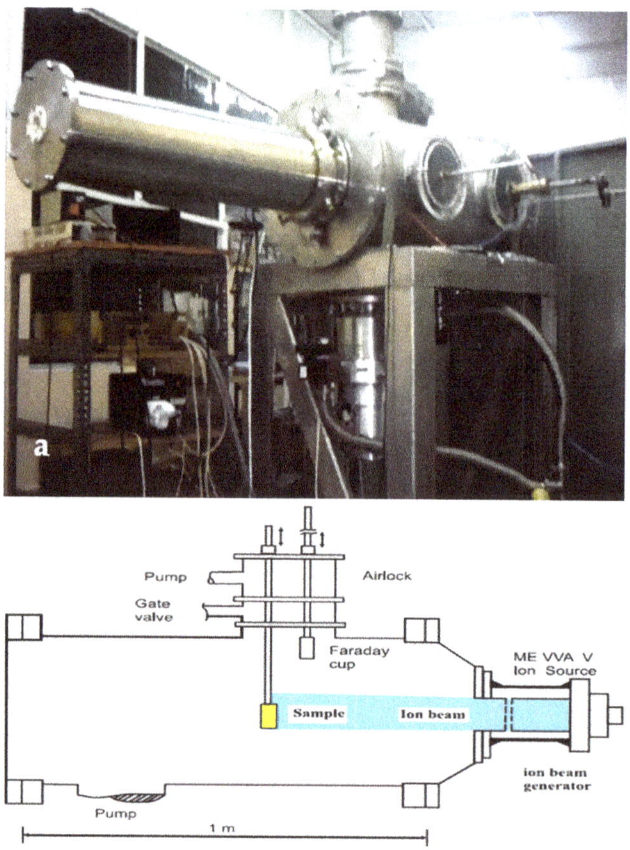

(b)

Figure 2. (**a**) MEVVA (metal vapor vacuum arc) ion implantation system. (**b**) Schematic illustration; Reprinted with permission from [14]; Copyright 2008 Elsevier.

In this research, the determination of the upper limits of the ion implantation dose and the pulse repetition rate was a serious challenge. These values were determined as a compromise between the ion dose levels and the goal of creating significantly improved surface features. In this research, we explain the results obtained for implantation doses between 5×10^{15} and 5×10^{16} ion/cm^2. Further increases in the ion dose caused surface overheating, and, therefore, they were not used. These doses were determined experimentally. In normal conditions, lower ion doses can only provide minor surface changes; however, in this study, ion doses between 5×10^{15} and 5×10^{16} ion/cm^2 were still satisfactory enough to create surface modification and, hence, characterization.

The ion source extraction voltage can be increased up to 110 kV. The substrate was placed in a vacuum chamber and bombarded by a beam of highly energized ions. Ag and Ti+O were implanted at an accelerating voltage of 30 kV using a MEVVA ion implantation source, with doses ranging from 5×10^{15} to 5×10^{16} ions per cm^2.

The ion penetration depth depends on both the energy of the ions and the atom density in the substrate. The effective parameters of MEVVA are ion species, energy, and ion dose and ion flux [21].

The features of the ion-implanted fabrics are shown in Table 1.

Table 1. Features of ion-implanted medical textiles.

No.	Antibacterial Agent (Cathode)	Fabric Type	Ion/cm^2 (Dose)
I1	Ag	Cotton	5×10^{15}
I2	Ag	Cotton	5×10^{16}
I3	Ag	Polyester	5×10^{16}
I4	Ag	Polyester	5×10^{15}
I5	TiO_2	Cotton	5×10^{16}
I6	TiO_2	Cotton	5×10^{15}
I7	TiO_2	Polyester	5×10^{16}
I8	TiO_2	Polyester	5×10^{15}

2.2. *Nanoparticle-Treated Samples*

There are many techniques in nanotechnology that provide antibacterial ability to textile surfaces [22]. However, only one technique was used in this study to provide antibacterial ability to the textile surfaces: the fixation of nanoparticles. This technique is also the most common method of providing desired abilities to textiles.

Table 2 presents the features of nanoparticle treated medical textiles.

Table 2. Features of nanoparticle-treated medical textiles.

Sample No.	Antibacterial Agent	Sample Type
N1	Ag	Cotton
N2	TiO_2	Polyester
N3	TiO_2	Cotton
N4	Ag	Cotton
N5	Ag	Polyester

The fixation of nanoparticles on the textile surfaces was achieved as follows: First, isopropanol was added to the ammonia aqueous solution of $AgNO_3$. Secondly, the 10×10 cm^2 piece of cotton or polyester textile was immersed in this solution and boiled for 1 h. Finally, the cotton or polyester sample was removed from the solution, washed with water, and sonicated several times before drying at 25 °C. At the end of this procedure, Ag nanoparticles were loaded onto the desired cotton or polyester textile. This procedure led to loadings of around 0.1 wt.% Ag/ cotton, as determined by elemental analysis [23]. The procedure led us to assume the same wt.% for Ag/polyester.

The TiO_2-loaded polyester/cotton was prepared by immersing the textile sample in a 5 g/L suspension of TiO_2 Degussa P-25, previously sonicated for 30 min in bidistilled water. The textile was then dried for 1 h and sonicated for 5 min at 75 °C to eliminate loosely bound particles. The loading of the cotton was around 0.1 wt.% TiO_2/cotton, as determined by elemental analysis [21].

2.3. Antibacterial Test Procedure

The antibacterial activities of ion-implanted and nanoparticle-treated samples were evaluated by AATCC Test Method 100-1999 [24] (Assessment of Antibacterial Finishes on Textile Materials). *Staphylococcus aureus* ATCC 6538 (a Gram-positive organism) was used as a test bacteria for antibacterial activity assessment. All bacterial experiments were implemented using nutrient broth/agar (NB/NA) medium consisting of peptone 5 g, beef extract 3 g, and distilled water 1 L. Fresh transplants of test bacteria were prepared from the stock culture before each experiment.

All swatch samples were cut into rectangles of 3 ± 0.1 cm size, and three pieces of fabrics were used for each repetition. All swatches were sterilized in an autoclave (121 °C 15 min) and were placed in sterile petri dishes aseptically. Fresh bacterial suspension was prepared using sterile saline solution (0.87% NaCl in distilled water) according to the McFarland standard (1–2 × 10^5 cfu/mL). All samples were inoculated with 400 mL of bacterial suspension, and petri dishes were closed with parafilm to prevent evaporation during the incubation at 37 °C for 24 h. After incubation, the swatches were transferred aseptically to a jar containing 40 mL sterile saline solution as a neutralizing solution and were shaken vigorously for one minute. One series of untreated swatches was used for contact time detection. Serial dilutions (10^{-1}, 10^{-2}) were made with saline solution and were plated (100 mL) on nutrient agar plates in duplicate. All plates were incubated at 37 °C for 24 h. Bacterial colonies were counted, and percentage reduction by the treated specimen was calculated by the following equation:

$$R = 100 \times [(B - A)/B]$$

R = (%) antibacterial efficiency reduction.

B = bacterial colonies belonging to untreated swatches, counted from petri dishes just after the inoculation.

A = bacterial colonies belonging to treated swatches (ion-implanted or nanoparticle treated), counted from petri dishes 24 h after the inoculation.

2.4. Laundry Procedure

Eight ion-implanted, five nanoparticle-treated, and two untreated fabrics were weighed as 58.004 g ≈ 58 g. According to the AATCC 124-2006 [25] Smoothness Appearance of Fabrics after Repeated Home Laundering test methodology, which states that 50 g of detergent must be used for 1 kg of fabric, it was necessary for us to use 3 g of detergent for the 58 g fabric in the experiment. However, to test the limit of antibacterial ability of the fabrics, 5 g of detergent was used for the 58 g fabric.

The fabrics were washed with 50 °C heated water. Washing progress was realized in a shaking incubator; 3 L of water was used for the 58.004 g fabric, according to AATCC 124-2006 test methodology. Fabrics were washed for 30 min with 290 rpm (rotation per minute). After that, the washed fabrics were rinsed in 35 °C heated water and, finally, they were dried at 40 °C heat for 25 min inside an incubator.

This test was repeated 30 times. After washing the textiles 30 times, the second round of antibacterial tests was administered to these textiles. Finally, after the second set of antibacterial test results was obtained, it was compared with the first round of antibacterial test results—which were done before the 30 washes—to evaluate which technique is more effective after 30 rounds of laundering.

2.5. Friction Resistance Test

The purpose of this test is to determine the appearance of textiles after induced abrasion. The measurements of the corrosion coefficient of textiles are very complex. Friction resistance is affected by many factors, such as inherent mechanical properties, dimensions, structures, construction, type, the kind of fabric, and amount of finishing materials added to the fabrics. In order to determine the friction resistance of fabrics, test samples are tested with an abrasive oscillatory cylinder under standard atmospheric conditions and predetermined loads (50 N) for a prescribed period (60 s) of time at room temperature. The frequency of oscillation was 0.1 Hz, with a Z-stroke length of 5 mm. These tests were realized according to the ASTM D4157-13 [26] standard test method for abrasion resistance of textiles. Untreated polyester textiles, with the features explained in Section 2.1, were utilized during the experiments [27].

2.6. UV Protection Factor Test

UV protection factor (UPF) tests are used to determine the ultraviolet radiation blocked by textiles intended to be used for UV protection [28]. A UPF value can simply be thought of as a time factor for a white-skinned person with skin protection compared to exposure without any protection. As an example, in the case where a person presents visible erythema after three minutes of exposure, fabric with a UPF of 30 extends that time to three minutes multiplied by the protection factor, i.e., 90 min, or, roughly, one and a half hours. The standard tests for computing UPF note that all measurements should be obtained from a UV–visible spectrophotometer that is equipped with an integrating sphere accessory. Fabric samples were placed at the transmittance port of the integrating sphere under moderate tension. Spectra of the fabric samples were collected from 280–400 nm using a special software (Shimadzu UV–vis–NIR spectroscopy, Kyoto, Japan). The software automates the determination of UPF, average UVA transmittance, and average UVB transmittance according to the AATCC 183:2014 [29] standard test method for transmittance or blocking of erythemally weighted ultraviolet radiation through fabrics. Untreated polyester textiles, with the same features explained in Section 2.1, were utilized during the experiments [30,31].

2.7. SEM Analysis

Scanning electron microscopy (Philips XL-305 FEG–SEM, Amsterdam, The Netherlands) was used to examine the surfaces of unimplanted and nanoparticle-treated textile surfaces. ×45 and ×100 magnification ratios were used for unimplanted cotton and polyester textiles, respectively. The ×3000 magnification ratio was used to evaluate the effects of nanoparticles on medical textiles. The details of the SEM analysis are written at the bottom of the SEM images.

3. Results and Discussion

In Figure 3, the results of all the antibacterial tests, before and after 30 washes, are presented in a graph.

It can easily be analyzed from this figure that the best results gained after 30 washes were the nanoparticle-treated cotton fabric and the ion-implanted polyester fabric.

3.1. Antibacterial Efficiency Results of Cotton Fabrics

The reason for this relies on the fact that cotton is a very absorbent fabric and nanoparticles were applied to the surfaces via liquid solutions containing nanoparticles [32]. When liquid solutions containing nanoparticles are applied to cotton fabric, the cotton fabric absorbs a lot of the liquid solution and gains many nanoparticles. As the solvent evaporates away, the nanoparticles start to accumulate on the surface and attach themselves to the surface with the help of physical bonds. This attachment increases as the roughness of the surface increases [33]. This phenomenon also increases the capability of cotton fabric to accumulate more nanoparticles since its roughness is much higher than polyester, as presented in the SEM pictures.

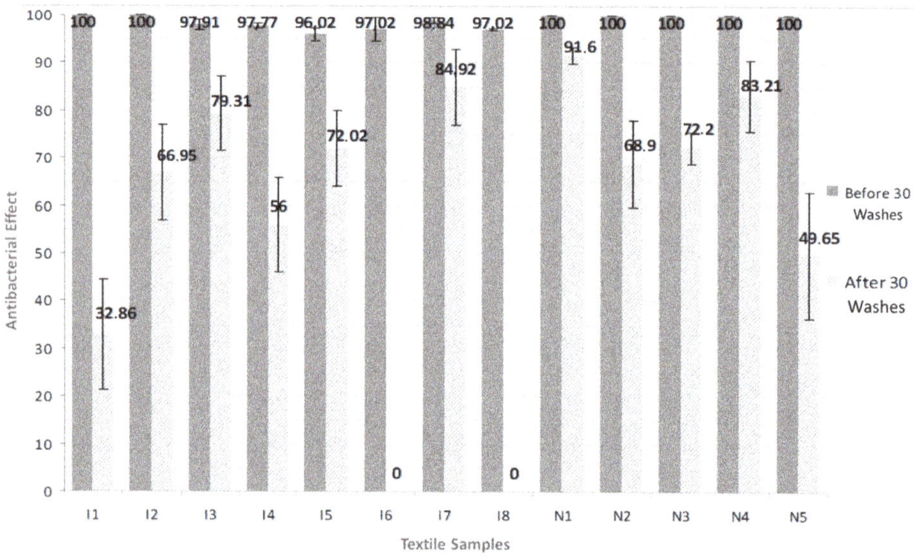

Figure 3. Antibacterial efficiency results, both before and after 30 washes.

On the other hand, polyester fabric is not absorbent as cotton fabric. The roughness of the surface of polyester fabric is also less than cotton fabric. Therefore, it cannot gain as many nanoparticles as cotton fabric can. Due to this reason, after 30 washes, nanoparticle-treated cotton fabrics have higher antibacterial efficiency results than both nanoparticle-treated polyester fabrics and ion-implanted cotton fabrics since they gained many more nanoparticles at the beginning.

As can easily be understood from the SEM micrograph in Figure 4a, because of surface morphology and the protruding ability of its surface, cotton fabric can hold a high number of nanoparticles, thus causing higher antibacterial efficiency.

(a)

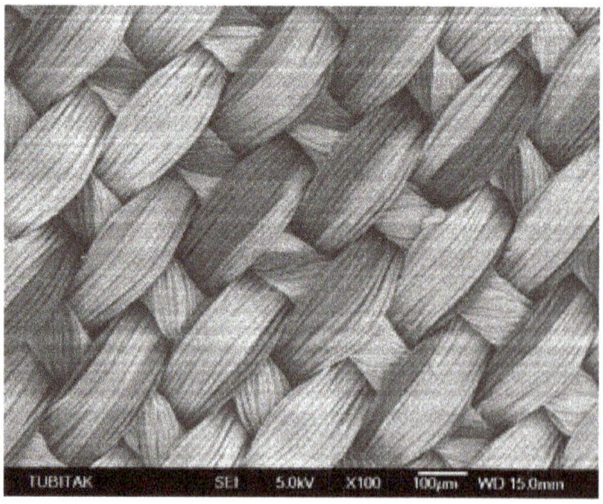

(b)

Figure 4. SEM micrographs of (**a**) cotton fabric and (**b**) polyester fabric.

In Figure 5, it can easily be analyzed that nanotextile-based cotton fabrics—N1, N4, and N3—resulted in the highest antibacterial efficiency rates, both after 30 washes and even before 30 washes. This is caused by surface morphology and the protruding ability of cotton fabric.

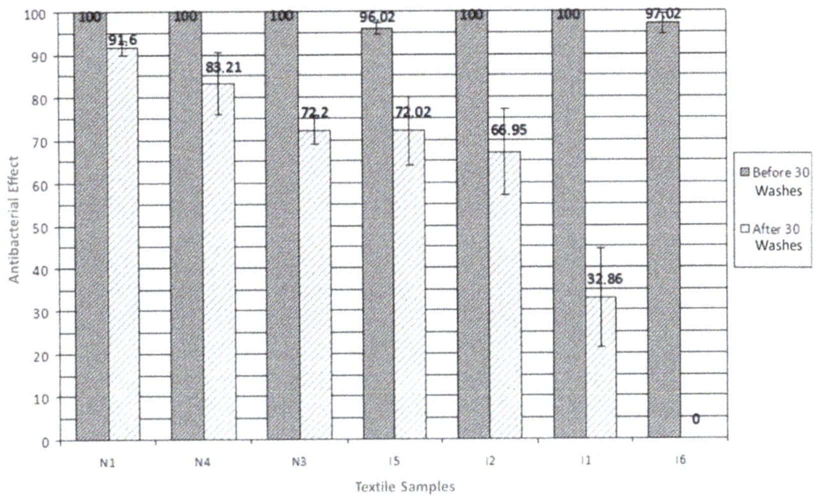

Figure 5. Cotton-based medical textile results gained from both nanoparticle-treated and ion-implanted fabrics.

Figure 6 presents the fibers of nanoparticle-treated medical textiles. As mentioned before, as the solvent evaporates, the nanoparticles accumulate on the surface and attach themselves to the surface and also to each other with the help of physical bonds. On the other hand, for ion implantation, depending on the material, the type of implanting element, and their kinetic energy, the depth of the ion implantation is usually between 0.1–2 µm. However, the specifications caused by the modification can be observed even at a depth of 50–100 µm. Therefore, the breathability and flexibility features of the textile materials do not change since the effect of surface modification is only limited to the surface and does not affect the bulk of the material.

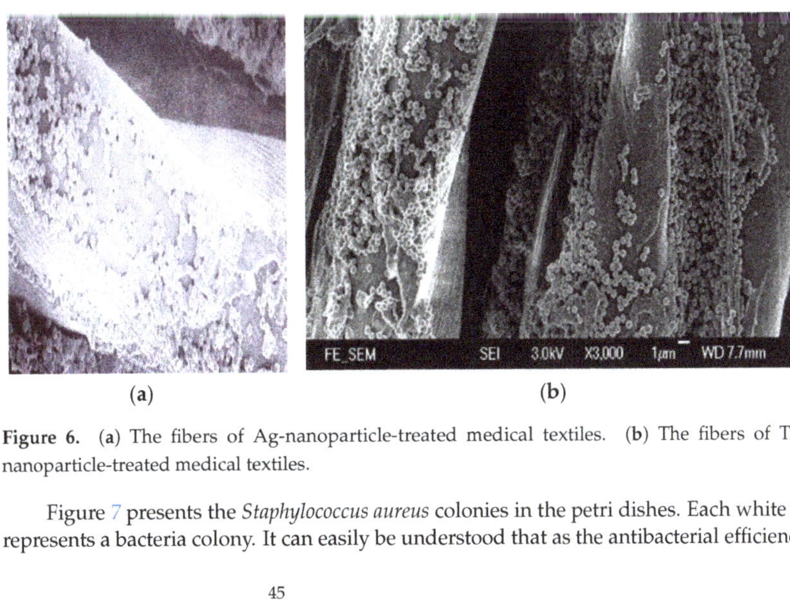

(a) (b)

Figure 6. (**a**) The fibers of Ag-nanoparticle-treated medical textiles. (**b**) The fibers of TiO_2-nanoparticle-treated medical textiles.

Figure 7 presents the *Staphylococcus aureus* colonies in the petri dishes. Each white dot represents a bacteria colony. It can easily be understood that as the antibacterial efficiencies decrease, the number of bacteria colonies increases. Before 30 washes, all improved fabrics presented almost 100% antibacterial efficiencies.

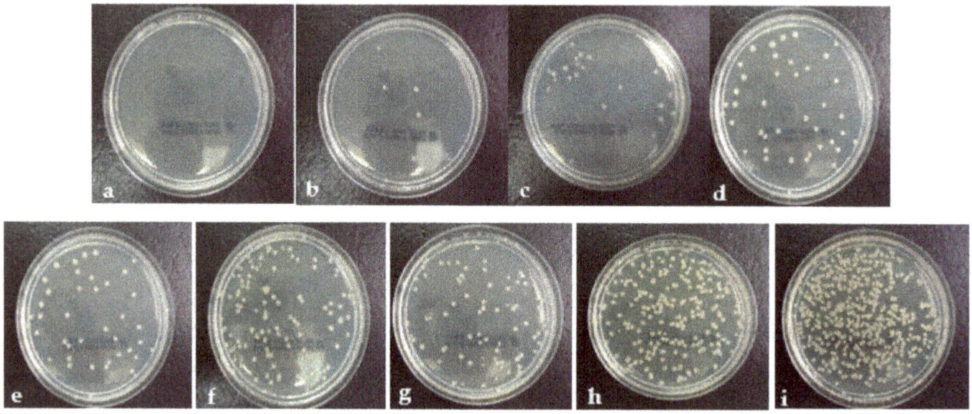

Figure 7. Antibacterial efficiency results before 30 washes: (**a**) 100%, (**b**) 96%, after 30 washes, (**c**) N1—91.60%, (**d**) N4—83.21%, (**e**) N3—72.20%, (**f**) I5—72.20%, (**g**) I2—66.95%, (**h**) I1—32.86%, and (**i**) I6—0%.

3.2. Results of Polyester Fabrics

It can easily be analyzed from Figure 8 that ion-implanted polyester textiles (especially I7 and I3) are superior to nanoparticle-treated polyester textiles. I7 and I3 also have some of the best results of the experiment.

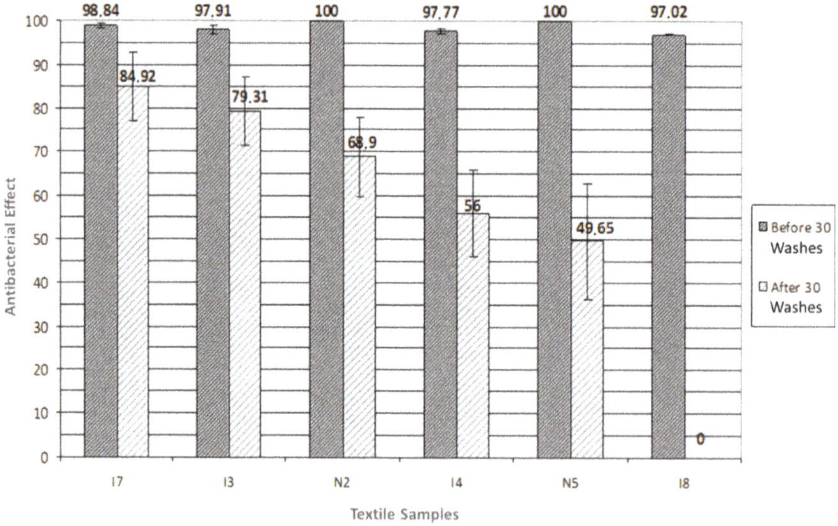

Figure 8. Polyester-based medical textile results gained from both nanoparticle-treated and ion-implanted polyesters. The polyester used in this project was composed of 75% polyester and 25% cotton fiber.

Similar to Figure 7, Figure 9 presents the *Staphylococcus aureus* colonies in the petri dishes. It can easily be understood that as the antibacterial efficiencies decrease, the number of bacteria colonies increases.

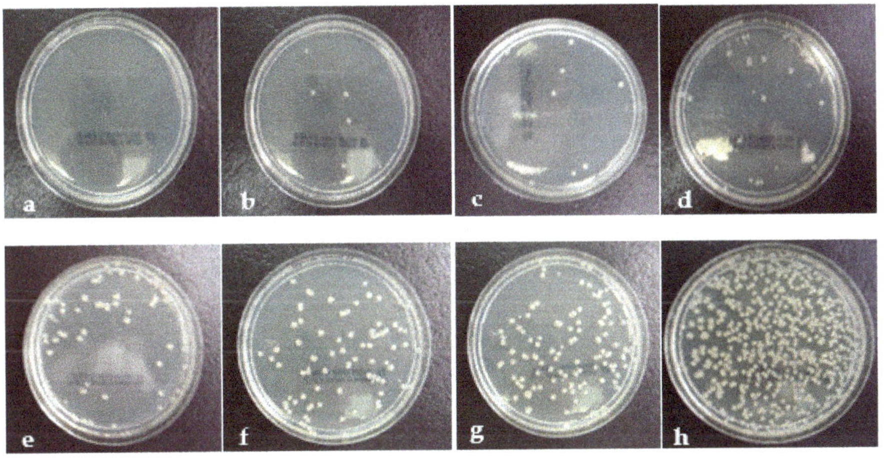

Figure 9. Antibacterial efficiency results before 30 washes: (**a**) 100%, (**b**) 96%, after 30 washes, (**c**) I7—84.92%, (**d**) I3—79.31%, (**e**) N2—68.90%, (**f**) I4—56%, (**g**) N5—49.65%, and (**h**) I8—0%.

The reason for antibacterial efficiency is that polyester fabrics are densely weaved textiles, and the surface area is much tighter than cotton fabric. As can be understood from the SEM micrograph in Figure 4b, polyester fabric has a smoother and denser weave than cotton. In this study, this had vital importance because smooth areas are better for ion beam implantation and result in higher antibacterial efficiency. Ion implantation technology is surface modification technology, and smoother areas at the textile surfaces are more suitable targets for ion beams. In this case, the ion beams collide with the surface at an inclined angle, and some of the ions are reflected or sputtered from the surface. This may affect the success of ion beam implantation. Therefore, it can be said that if the roughness increases, the success rate of ion implantation decreases. This problem is presented in Figure 10. Based on this fact, even after 30 washes, ion-implanted polyester fabrics have more antibacterial efficiency results than both nanoparticle-treated polyester fabrics and ion-implanted cotton fabrics since they gained many more ion beams at the beginning due to their smooth surface structure.

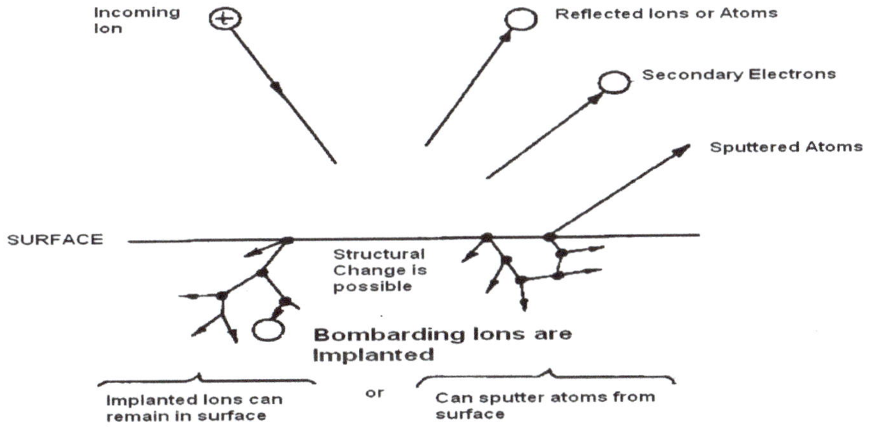

Figure 10. Ion implantation at an inclined angle.

3.3. *In the Case of Higher Ion Doses*

Table 3 describes the ion/cm^2 based arrangement of features of ion implanted medical textles.

Table 3. Ion/cm^2-based arrangement of features of ion-implanted medical textiles.

No.	Ion/cm^2 (Dose) (Cathode)	Fabric Type	Antibacterial Agent
I1	5×10^{15}	Cotton	Ag
I4	5×10^{15}	Polyester	Ag
I6	5×10^{15}	Cotton	TiO_2
I8	5×10^{15}	Polyester	TiO_2
I5	5×10^{16}	Cotton	TiO_2
I3	5×10^{16}	Polyester	Ag
I7	5×10^{16}	Polyester	TiO_2
I2	5×10^{16}	Cotton	Ag

The ion doses used in the project were between 5×10^{15} and 5×10^{16} dose/cm^2. According to the antibacterial efficiency results, when ion doses were increased, antibacterial efficiency results also increased significantly. If higher doses were used in the research, antibacterial efficiency results would have been much higher before and, especially, after 30 washes. However, a dose rate of over 1×10^{17} dose/cm^2 cannot be applied to the fabrics due to overheating.

Figure 11 presents a comparison of similar ion treatments. It can easily be analyzed that when the ion dose increased ten times, antibacterial efficiency results increased over 30%. It is not unrealistic to estimate that if a two-time higher ion dose (1×10^{17} ion/cm^2) was used, it would have resulted in much better antibacterial efficiency scores. A logarithmic line is also inserted into the figure in order to make it easier to evaluate the trend caused by the increase in ion implantation dosage.

In addition, Figure 11c,d also present that TiO_2 is not as resistive as Ag against washing. No antibacterial ability was present in 5×10^{15} ion/cm^2 TiO_2 ion-implanted cotton and polyester textiles after 30 washes. The reason is probably that the laundry process breaks the bonds between titanium and oxygen thus removing the antibacterial ability of TiO_2.

3.4. *Friction Coefficiency Results of 5×10^{15} Ion/cm^2 Titanium Ion-Implanted Polyester Fabric*

Beyond antibacterial ability, friction and UV protection capabilities were also analyzed in order to understand the other contributions of ion implantation into textile surfaces. Figure 12 shows the pictures of untreated and Ti-implanted polyester textiles. The marks on the surfaces are caused by the friction resistance test, explained in Section 2.5. Due to implantation, the color of the polyester turned from yellow to brown.

As a result, the friction coefficient decreased by 50% of its untreated first value when 5×10^{15} ion/cm^2 Ti was implanted into the polyester surfaces. Figure 13 presents the friction coefficient results of both textiles. This result implies that the implanted textiles may resist wear and corrosion much longer than untreated textiles; therefore, they may be used for decades longer than untreated ones.

Figure 14 provides much deeper information for this phenomenon. SEM micrographs of untreated and 5×10^{15} ion/cm^2 Ti-implanted samples were obtained and compared. It is obviously understood that ion implantation flattens the peaks, fills the trenches, and leads to smoother surfaces. Therefore, due to the nature of ion implantation, ion-implanted surfaces develop friction resistance automatically.

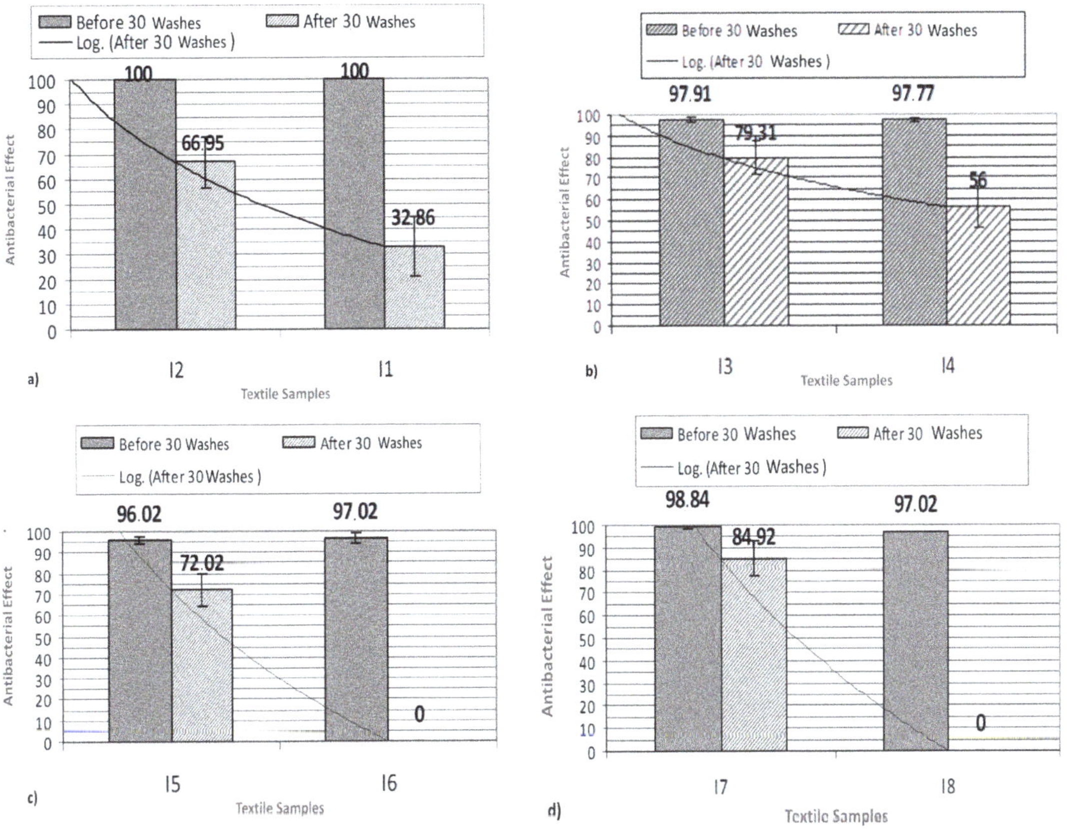

Figure 11. Comparison of (**a**) I2 (cotton, Ag, 5×10^{16} ion/cm^2) and I1 (cotton, Ag, 5×10^{15} ion/cm^2), (**b**) I3 (polyester, Ag, 5×10^{16} ion/cm^2) and I4 (polyester, Ag, 5×10^{15} ion/cm^2), (**c**) I5 (cotton, TiO_2, 5×10^{16} ion/cm^2) and I6 (cotton, TiO_2, 5×10^{15} ion/cm^2), and (**d**) I7 (polyester, TiO_2, 5×10^{16} ion/cm^2) and I8 (polyester, TiO_2, 5×10^{15} ion/cm^2).

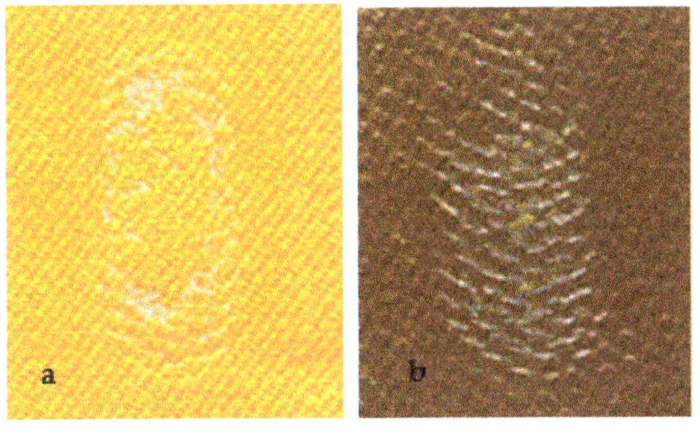

Figure 12. Pictures of (**a**) untreated and (**b**) 5×10^{15} ion/cm^2 Ti-implanted polyester fabric after the same number of cycles in an abrasive oscillatory cylinder.

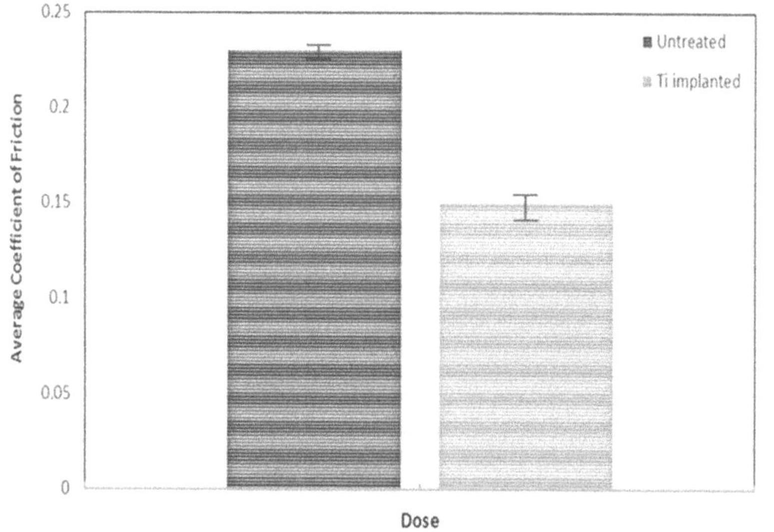

Figure 13. Comparison of the average coefficient of friction values of untreated and 5×10^{15} ion/cm^2 Ti-implanted polyester fabrics.

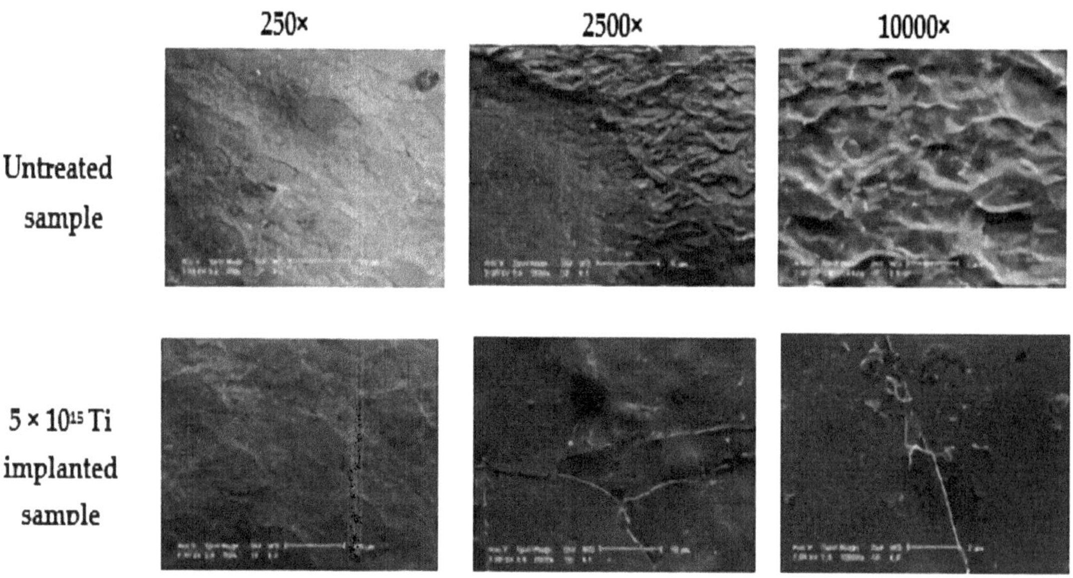

Figure 14. SEM micrographs of untreated and 5×10^{15} ion/cm^2 Ti-implanted samples.

3.5. UV Protection Factor of 5×10^{15} Ion/cm^2 Ag Ion-Implanted Polyester Fabric

After UV protection factor (UPF) tests were carried out on 5×10^{15} ion/cm^2 Ti and 5×10^{15} ion/cm^2 Ag ion-implanted samples, the results were compared with an untreated sample. As a result, it is understood that while Ti-implanted samples increased the UPF by more than two times, Ag-implanted samples increased UPF by more than six times. Therefore, Ag-implanted samples can be categorized as samples with a very good protection factor. It will be a strong suggestion to claim that if the ion dose is increased to

1×10^{17} ion/cm^2, the UPF would be higher. The UPF classification is shown in Table 4, and the results are presented in Figure 15.

Table 4. UV protection factor classification.

UV Protection Factor	Protection Standard
15–24	Good Protection
25–39	Very Good Protection
40–50, 50+	Excellent Protection

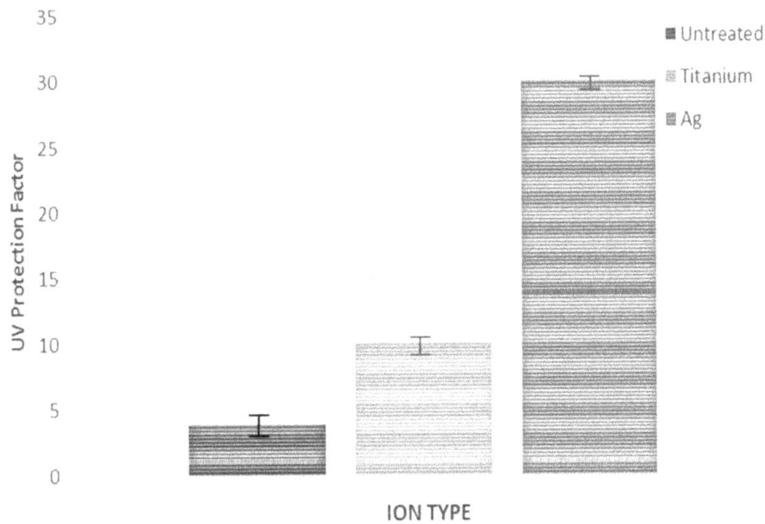

Figure 15. Comparison of UV protection factors of untreated, 5×10^{15} ion/cm^2 Ti ion-implanted, and 5×10^{15} ion/cm^2 Ag ion-implanted samples.

4. Conclusions

In this research, first, a group of polyester- and cotton-based medical textiles were treated with Ag and TiO_2 ion beams in different doses. Then, another group of polyester- and cotton-based medical textiles were treated with nanoparticles contained in liquid solutions. The antibacterial efficiencies of the improved fabrics were compared before and after 30 washes in order to evaluate their resistance against laundering and which technique enables antibacterial agents to bind to fabric surfaces strongly.

At the end of the experiment, it was understood that while nanoparticles are superior for cotton fabric, ion implantation is more successful with polyester fabrics. Overall, it can be said that both techniques are equal in creating antibacterial effects. However, some researchers claim that there may be a toxicity problem caused by nanoparticles. In case of a discovery of serious toxicity problems with nanoparticles that are used in nanotextiles, ion implantation technology, which has no known toxicity problem, is ready to serve the textile industry.

In addition to antibacterial efficiency, it is proven that Ag and Ti+O implantation have serious effects on friction resistance and UV protection. Therefore, in conclusion, if the ion doses and textile features are optimized or other ion implantation techniques are utilized, it will not be just a dream to fabricate a "smart textile" with infinite antibacterial efficiency, a very low friction resistance that enables textiles to endure for generations against corrosion, and very high UV protection capability that reduces cancer risk effectively. Additionally, water permeability, wrinkle recovery, and radiation absorbance features of textile surfaces may be enhanced when Ag and Ti+O ions are implanted. Also medical textiles with infinite antibacterial ability will have capability to reduce the global medical waste amount significantly [34]. This is another research idea that need to be focused. All these ideas may be considered in future studies.

Funding: This research received no external funding.

Institutional Review Board Statement: Not applicable.

Informed Consent Statement: Not applicable.

Data Availability Statement: The data presented in this study are available within the article.

Acknowledgments: I would like to thank Mustafa Ahmet ÖZTARHAN for all the effort, supervision, equipment, and SEM images he has provided for this research. Unfortunately, during the preparations of the manuscript, he passed away. May he rest in peace forever.

Conflicts of Interest: The authors declare no conflict of interest.

References

1. Mejía, M.L.; Zapata, J.; Cuesta, D.P.; Ortiz, I.C.; Botero, L.E.; Galeano, B.J.; Hoyos, L.M. Properties of antibacterial nano textile for use in hospital environments. *Revis. Ing. Biomed.* **2017**, *11*, 13–19. [CrossRef]
2. Morais, D.S.; Guedes, R.M. Antimicrobial approaches for textiles: From research to market. *Materials* **2016**, *9*, 498. [CrossRef] [PubMed]
3. Gokarneshan, N.; Chandrasekar, P.T.; Suvitha, L. Advances in nanotextile finishes—An approach towards sustainability. In *Textiles and Clothing Sustainability*; Muthu, S., Ed.; Springer: Singapore, 2017; pp. 1–56.
4. Islam, S.; Shabbir, M.; Mohammad, F. Insights into the functional finishing of textile materials using nanotechnology. In *Textiles and Clothing Sustainability*; Muthu, S., Ed.; Springer: Singapore, 2017; pp. 97–115.
5. Krce, L.; Šprung, M.; Maravić, A.; Umek, P.; Salamon, K.; Krstulović, N.; Aviani, I. Bacteria exposed to silver nanoparticles synthesized by laser ablation in water: Modelling *E. coli* growth and inactivation. *Materials* **2020**, *13*, 653. [CrossRef] [PubMed]
6. Li, J.; He, J.; Huang, Y. Role of alginate in antibacterial finishing of textiles. *Int. J. Biol. Macromol.* **2017**, *94*, 466–473. [CrossRef] [PubMed]
7. Turakhia, B.; Divakara, M.B.; Santosh, M.S.; Shah, S. Green synthesis of copper oxide nanoparticles: A promising approach in the development of antibacterial textiles. *J. Coat. Technol. Res.* **2020**, *17*, 531–540. [CrossRef]
8. Zemljič, L.F.; Peršin, Z.; Šauperl, O.; Rudolf, A.; Kostić, M. Medical textiles based on viscose rayon fabrics coated with chitosan-encapsulated iodine: Antibacterial and antioxidant properties. *Text. Res. J.* **2018**, *88*, 2519–2531.
9. Fouda, A.; Saad, E.L.; Salem, S.S.; Shaheen, T.I. In-vitro cytotoxicity, antibacterial, and UV protection properties of the biosynthesized zinc oxide nanoparticles for medical textile applications. *Microb. Pathog.* **2018**, *125*, 252–261. [CrossRef]
10. Ouadil, B.; Amadine, O.; Essamlali, Y.; Cherkaoui, O.; Zahouily, M. A new route for the preparation of hydrophobic and antibacterial textiles fabrics using Ag-loaded graphene nanocomposite. *Colloids Surf.* **2019**, *579*, 123713. [CrossRef]
11. Edgetech. Ion Implantation. Available online: https://www.edge-techind.com/category/Ion-implantation-44-1.html (accessed on 17 January 2021).
12. IBAD Lab. IBAD Laboratory. Available online: https://fyzika.fs.cvut.cz/laboratorIBADEN.php (accessed on 17 January 2021).
13. Öktem, T.; Özdogan, E.; Namligöz, S.E.; Öztarhan, A.; Tek, Z.; Tarakçioglu, I.; Karaaslan, A. Investigating the applicability of metal ion implantation technique (MEVVA) to textile surfaces. *Text. Res. J.* **2006**, *76*, 32–40.
14. Öktem, T.; Tarakcıoğlu, I.; Özdoğan, E.; Öztarhan, A.; Namlıgöz, E.S.; Karaaslan, A.; Tek, Z. Modification of friction and wear properties of PET membrane fabrics by MEVVA ion implantation. *Mat. Chem. Phys.* **2008**, *108*, 208–213.
15. Nikolaev, A.G.; Yushkov, G.Y.; Oks, E.M.; Oztarhan, A.; Akpek, A.; Hames-Kocabas, E.; Urkac, E.S.; Brown, I.G. Modification of anti-bacterial surface properties of textile polymers by vacuum arc ion source implantation. *Appl. Surf. Sci.* **2014**, *310*, 51–55. [CrossRef]
16. Anna, K.; Mateusz, S.; Eva., K. Similarities and differences between silver ions and silver in nanoforms as antibacterial agents. *Int. J. Mol. Sci.* **2018**, *19*, 444.
17. Alavi, M.; Rai, M. Recent advances in antibacterial applications of metal nanoparticles (MNPs) and metal nanocomposites (MNCs) against multidrug-resistant (MDR) bacteria. *Expert Rev. Anti-Infect.* **2019**, *17*, 419–428. [CrossRef]
18. Ward, L.P.; Purushotham, K.P.; Manory, R.R. MEVVA ion implantation of TiCN coatings; structural and tribological properties. *Nucl. Instrum. Methods Phys. Res. Sect.* **2019**, *449*, 40–48. [CrossRef]
19. Wang, C.C.; Chen, C.C. Physical properties of crosslinked cellulose catalyzed with nano titanium dioxide. *J. Appl. Polym. Sci.* **2005**, *97*, 2450–2456. [CrossRef]
20. Vaishya, R.; Agarwal, A.K.; Tiwari, M.; Vaish, A.; Vijay, V.; Nigam, Y. Medical textiles in orthopedics: An overview. *J. Clin. Orthop. Trauma* **2018**, *9*, S26–S33. [CrossRef] [PubMed]

21. Ward, L.P.; Purushotham, K.P.; Manory, R.R. Nuclear instruments and methods in physics research B studies on the surface modification of TiN coatings using MEVVA ion implantation with selected metallic species. *Nucl. Insturm. Methods Phys. Res.* **2016**, *368*, 37–44. [CrossRef]
22. Dastjerdi, R.; Montazer, M. A review on the application of inorganic nano-structured materials in the modification of textiles: Focus on anti-microbial properties. *Coll. Surf.* **2010**, *79*, 5–18. [CrossRef]
23. Yuranova, T.; Rincon, A.G.; Pulgarin, C.; Laub, D.; Xantopoulos, N.; Mathieu, H.J.; Kiwi, J. Performance and characterization of Ag-cotton and Ag/TiO_2 loaded textiles during the abatement of *E. coli*. *J. Photochem. Photobiol.* **2006**, *181*, 363–369. [CrossRef]
24. *AATCC Test Method 100-1999, Assessment of Antibacterial Finishes on Textile Materials*; AATCC: Research Triangle Park, NC, USA, 1999.
25. *AATCC 124-2006, Test Method Appearance of Fabrics After Repeated Home Laundering*; AATCC: Research Triangle Park, NC, USA, 2006.
26. *ASTM D4157-13, Standard Test Method for Abrasion Resistance of Textile Fabrics (Oscillatory Cylinder Method)*; ASTM International: West Conshohocken, PA, USA, 2013.
27. Öztarhan, A.; Akpek, A.; Oks, E.; Nikolaev, A. Modifying medical textiles with antibacterial and friction resistance abilities by an alternative nanotextile technology called ion implantation technique. In Proceedings of the 2010 15th National Biomedical Engineering Meeting, Antalya, Turkey, 21–24 April 2010; pp. 1–4.
28. Menter, J.M.; Hatch, K.L. Clothing as solar radiation protection. *Curr. Probl. Derm.* **2003**, *31*, 50–63.
29. *AATCC 183:2014, Test Method for Transmittance or Blocking of Erythemally Weighted Ultraviolet Radiation through Fabrics*; AATCC: Research Triangle Park, NC, USA, 2014.
30. Öztarhan, A.; Akpek, A.; Oks, E.; Nikolaev, A. Modifying textiles with antibacterial effect, friction resistance, UV protection and electrostatic charge decay abilities by an alternative nanotextile technology called MEVVA ion implantation technique. In Proceedings of the VIth Nanoscience and Nanotechnology Conference (NANOTR), Izmir, Turkey, 15–18 June 2010.
31. Nikolaev, A.G.; Savkin, K.P.; Yushkov, G.Y.; Oks, E.M.; Oztarhan, A.; Akpek, A.; Cireli, I. Modification of the textile materials by vacuum arc ion source implantation. In Proceedings of the 10th International Conference on Modification of Materials with Particle Beams and Plasma Flows (10th CMM), Tomsk, Rusya, 19–24 September 2010; pp. 401–404.
32. Buras, E.M., Jr.; Goldthwait, C.F.; Kraemer, R.M. Measurement and theory of absorbency of cotton fabrics. *Text. Res. J.* **1950**, *20*, 239–248. [CrossRef]
33. Nayfeh, M.H. (Ed.) Delivery of nanoparticles on surfaces. In *Fundamentals and Applications of Nano Silicon in Plasmonics and Fullerines: Current and Future Trends*; Elsevier: Amsterdam, The Netherlands, 2018; pp. 341–362.
34. Koçak, O.; Kurtuldu, H.; Akpek, A.; Koçoğlu, A.; Eroğul, O. A medical waste management model for public private partnership hospitals. In Proceedings of the 2016 Medical Technologies National Congress (TIPTEKNO), Antalya, Turkey, 27–29 October 2016; pp. 1–4.

coatings

Article

An Investigation on the Thermal and Solar Properties of Graphene-Coated Polyester Fabrics

Gizem Manasoglu, Rumeysa Celen *, Mehmet Kanik and Yusuf Ulcay

Textile Engineering Department, Bursa Uludag University, 16240 Bursa, Turkey; gmanas@uludag.edu.tr (G.M.); mekanik@uludag.edu.tr (M.K.); ulcay@uludag.edu.tr (Y.U.)
* Correspondence: rumeysa@uludag.edu.tr

Abstract: In this study, coatings were made with graphene nanopowder in two different thicknesses (0.1 and 0.5 mm) at three different concentrations (50, 100 and 150 g/kg) on polyester woven fabrics. The effects of the coating thickness and graphene concentration were examined with optical and scanning electron microscopy (SEM) images. The thermal stability properties of the samples were also evaluated by differential scanning calorimetry (DSC) and thermal gravimetric analysis (TGA). Thermal conductivity was evaluated with two different principles: contact and radiant heat transfer, according to JIS R 2618 and EN ISO 6942, respectively. Solar measurements were performed with a Shimadzu UV-3600 Plus spectrophotometer. The graphene coating improved the thermal stability of the polyester fabrics. The solar absorbance value increased by 80% compared to reference fabrics, and reached approximately 90%. One of the important results was that the thermal conductivity coefficient increased by 87% and 262% for the two coating thicknesses, respectively.

Keywords: graphene; polyester; thermal conductivity; heat transmission; near-infrared; reflectance; absorbance

Citation: Manasoglu, G.; Celen, R.; Kanik, M.; Ulcay, Y. An Investigation on the Thermal and Solar Properties of Graphene-Coated Polyester Fabrics. *Coatings* **2021**, *11*, 125. https://doi.org/10.3390/coatings11020125

Received: 11 December 2020
Accepted: 19 January 2021
Published: 23 January 2021

1. Introduction

Today, coating is an essential requirement of many technical textiles, which would not be capable of satisfying the strict requirements on functionality and standards without the application of special raw materials [1]. The coating enhances and extends the range of functional performance properties of textiles, and the use of these techniques is growing rapidly as the applications for technical textiles become more diverse [2]. The focus in textile coating has been on the development of new materials that provide specific characteristics such as electromagnetic shielding, electrical conductivity, thermal insulation, thermal conductivity, and sound absorption, etc. [2]. Particularly, nano-size fillers such as clay, metal oxides, carbon black have been used in composite materials. Due to their large surface area, they have a better interaction with polymer matrices and have better performance and new market interest [3].

Although graphene was discovered in 1859, there has been an explosion in the exploration of the material since 2004, with the study of A. K. Geim and K. S. Novoselov. Graphene is an excellent material and has been considered a promising candidate for the post-silicon age [4]. Graphene is the first truly 2D crystal ever observed in nature [5]. Graphene is a single layer of carbon atoms densely packed in a honeycomb lattice [6]. The term graphene is typically applied to a single graphite layer, although common references also exist to bilayer or trilayer graphene [7]. The definition of graphene nano platelets (GNP) is still under review and is subject to change. Nevertheless, this material can broadly be defined as graphene with a typical thickness of 1–3 nm and lateral dimensions ranging from approximately 100 nm to 100 μm [8]. Graphene has unique electronic, optical, thermal, mechanical properties, a stable chemical nature, and low density. Therefore, it differs from most conventional three-dimensional materials and has extensive usability [9,10].

Energy conservation and environmental protection have become the most important issues for humanity. Heat transfer rate is a vital factor in determining the efficiency of thermal energy storage system. Enhancing thermal conductivity is an effective approach to improving thermal energy storage systems [10]. Efficient heat management systems have become extremely important in various fields, such as electronic, optoelectronic, and thermoelectric applications. In this regard, the development of materials with high thermal conductivity is urgently needed [11]. Carbon allotropes and their derivatives occupy a unique place in terms of their ability to conduct heat. The thermal conductivity of carbon materials at room temperature spans an extraordinarily large range—of over five orders of magnitude—from the lowest in amorphous carbons, to the highest in graphene and carbon nanotubes [12].

The high thermal conductivity of graphene can be used within composites to create thermal interface materials for electronics, thermal sensors and energy management systems, optical devices, optoelectronics, sensors/detectors, and composites/barriers. The thermal conductivity varies with the size of graphene flake and markedly decreases with the number of carbon layers, even in clean suspended graphene [8]. Some researchers have studied coated textiles with graphene or graphene derivatives to enhance their thermal conductivity property. For example, Gan et al. [13] coated cotton fabric using the wet coating technique. The graphene nano-ribbon coating improved the thermal stability of the cotton fabric. In the study of Abbas et al. [14], graphene, multi-wall carbon nanotube (MWCNT) and fine boron nitride (BN) particles were separately applied with a resin onto cotton fabric, and the effect of the thin composite coatings on the thermal conductive property was examined. The study has indicated that the graphene-coated fabric showed the best improvement in the thermal conductivity. Gunesekara et al. [15] investigated the possibility of using graphene oxide-coated fabric for thermal conductive purposes. The thermal conductivity of coated knitted fabric was improved compared to the untreated control sample. In the study of Hu et al. [9], graphene/PU coatings were applied to cotton fabrics according to the pad-dry-cure process. The results showed that the increase in graphene-coated weight enhanced the thermo-physical properties (thermal conductivity and thermal resistance) of cotton fabrics by nearly 30%.

In this study, graphene nano platelets were applied to the polyester woven fabric by the knife-over-roll coating process. Fabrics were coated at three different graphene concentrations and two different thicknesses for each concentration rate. The thermal conductivity of graphene-coated fabrics has been evaluated, for the first time, by two methods with different principles (contact and radiant), supported by solar measurements. Additionally, DSC and TGA analyses were performed to determine the thermal stability of samples.

2. Materials and Methods

2.1. Materials

Pre-treated (desized and thermofixed) polyester fabrics, which were used in the experiments were supplied by the DKC Technical Coating Company (Bursa, Turkey). The structural properties of the base fabric are given in Table 1.

Table 1. Properties of the base fabric.

Property	Warp	Weft
Yarn Type	Texturized polyester	Texturized polyester
Yarn Count (denier)	300/72	300/72
Yarn Density (thread/cm)	30	18
Yarn Crimp (%)	1.16	0.30
Fabric Mass per Unit Area (g/m^2)	169	
Fabric Thickness (mm)	0.34	

The coating chemicals such as the binder, thickener, fixation agent, anti-foam agent, and dispergator were supplied from Rudolf Duraner (Bursa, Turkey), and their properties are given in Table 2.

Table 2. Properties of coating chemicals.

Chemical	Property
Binder	Acrylic binder, anionic/nonionic
Synthetic thickener	Neutralized polyacrylate, anionic
Fixation agent	The butanone oxime-free blocked isocyanate-based cross-linking agent, anionic
Anti-foam agent	Hydrocarbons, ethoxylated fatty acids and silicic acid combination, nonionic
Dispergator	Anionic
Ammonia	25% liquid

Graphene nanoplatelets powder, supplied by the Grafen Chemical Industries (Ankara, Turkey), was used as a filler material and its properties are given in Table 3.

Table 3. Properties of graphene.

Property	Value
Purity (%)	96–99
Surface area(m^2/g)	13–15
Thickness (nm)	50–100
Diameter (µm)	~5

2.2. Methods

2.2.1. Preparing the Coating Paste

Firstly, the stock paste was prepared with mixing binder (350 g) and reverse osmosis water (597 g). Ammonia (5 g) was added to this mixture. Then the other chemicals; fixation agent (25 g), antifoam agent (6 g) and synthetic thickener (17 g) were added to the paste. Viscosity measurements were performed with Brookfield RVT analog viscometer (Middleboro, MA, USA). Viscosity value of the stock paste varied in the range of 4000 ± 200 cP. Coating pastes which had 7000 ± 200 cP viscosity were prepared by graphene powder at three different concentration. Graphene powders were dispersed in a certain amount of water and dispergator mixture and they were added to the paste. The recipe of coating paste is given in Table 4.

Table 4. Recipe of coating paste.

Stock Paste (g)	800
Graphene (g)	X (50, 100, 150)
Dispergator (g)	2
Anti-foam agent (g)	14
Thickener/Water (g)	Y
Total (g)	1000

2.2.2. Coating

The coating process was performed according to the knife-over-roll technique on an Atac GK 40 RKL laboratory type coating machine (Figure 1). The distance between the knife and fabric was arranged as 0.1 and 0.5 mm for three concentration rates (50, 100 and 150 g/kg) to examine the effect of different coating thicknesses. Coated fabrics were dried at 100 °C for 10 min. They were fixed at 160 °C for 3 min in a Rapid H-TS-3 laboratory type steamer. Fabric codes are given in Table 5. To investigate the sole effect of graphene on measurements, coated samples were compared with the blind coatings (R2.1 and R2.5).

Figure 1. Coating process on Atac GK 40 RKL.

Table 5. Fabric codes.

Code	Definition
R2.1	Stock paste-coated reference fabric which has no filler (graphene) at 0.1 mm
R2.5	Stock paste-coated reference fabric which has no filler (graphene) at 0.5 mm
GR50.1	Coated fabric with 50 g/kg at 0.1 mm
GR50.5	Coated fabric with 50 g/kg at 0.5 mm
GR100.1	Coated fabric with 100 g/kg at 0.1 mm
GR100.5	Coated fabric with 100 g/kg at 0.5 mm
GR150.1	Coated fabric with 150 g/kg at 0.1 mm
GR150.5	Coated fabric with 150 g/kg at 0.5 mm

2.2.3. Test and Characterization

Thickness and mass per unit are measurements of fabrics were made according to TS 7128 EN ISO 5084 [16] and TS 251 standards [17], respectively. Equation (1) was used to determine the amount of substance transferred:

$$W_3 = W_2 - W_1 \quad (1)$$

where W_1 (g/m^2) is the weight of uncoated base fabric, W_2 (g/m^2) is the weight of the coated fabric, and W_3 (g/m^2) is the add-on amount (g/m^2).

Optical images (both front and reverse sides) of reference fabrics (R2.1 and R2.5) and graphene-coated fabrics were taken with an Mshot Digital Microscope Camera MS60 (Guangzhou, China) with an objective of 30× magnification rate. SEM images of reference fabrics (R2.1 and R2.5) and graphene-coated fabrics were taken with a Carl Zeiss/Gemini 301 scanning electron microscope (Jena, Germany). A gold coating was applied to the samples before analysis, and the acceleration voltage was 10 kV. The magnification rates were chosen as 100× and 150×.

The DSC analyses were performed to determine the melting temperature (T_m) of the reference (R2.1 and R2.5) and graphene-coated samples using a heat flux TA/Discovery DSC 251 instrument with a heating rate of 10 °C/min from 50 to 350 °C under a nitrogen environment. The amount of test specimen was about 10 mg.

The TGA analyses were performed by a TA/SDT 650 thermo-gravimetric analyzer (New Castle, DE, USA). TGA and differential thermal gravimetric (DTG) curves, the decomposition temperatures, and total mass loss values were obtained. The amount of test specimen was about 10 mg. The initial and the final temperatures were 30 and 600 °C respectively. The heating rate was 20 °C/min, and the gas type was nitrogen.

Thermal properties of fabrics were evaluated according to two different standards. The device (MEK model, Figure 2) that measures the radiant heat transfer according to EN

ISO 6942 standard [18] was produced by Hasimoglu Mamatlar Machinery (Bursa, Turkey). In this study, method B (one of the methods related as standard) was used to estimate the heat transmission factor (TF, %) and the transmitted heat flux density (Q_C, kW/m^2). These parameters were evaluated by measuring the time (s) necessary to obtain a temperature increase of 12 ± 0.1 and 24 ± 0.2 °C (t_{12} and t_{24}). The transmitted heat flux density Q_C (kW/m^2) was calculated according to Equation (2):

$$Q_C = \frac{M \times C_P \times 12}{A \times (t_{24} - t_{12})} \tag{2}$$

where M is the copper plate (kg), C_P is the specific heat of copper (kJ/kg·°C), $12/(t_{24}-t_{12})$ is the mean rate of rising of the calorimeter temperature in °C/s, in the region between a 12 °C and a 24 °C rise, and A is the area of the copper plate (m^2).

The heat transmission factor TF (Q_0), for the incident heat flux density level Q_0 (kW/m^2), was given by Equation (3):

$$TF(Q_0) = \frac{Q_C}{Q_0} \tag{3}$$

The incident heat flux density (Q_0) was chosen as 5 kW/m^2 (low level) according to the standard. Silicon carbide heating rods were used to provide the required heat flux. The samples were placed in a specimen holder (frame) and they were exposed to radiant heat. Exposure duration was controlled using a manually operated shutter. The time taken for the temperature to rise by 12 ± 0.1 and 24 ± 0.2 °C in the copper plate calorimeter were measured. All samples were conditioned at a temperature of 20 ± 2 °C and (65 ± 2) % RH (relative humidity) before testing.

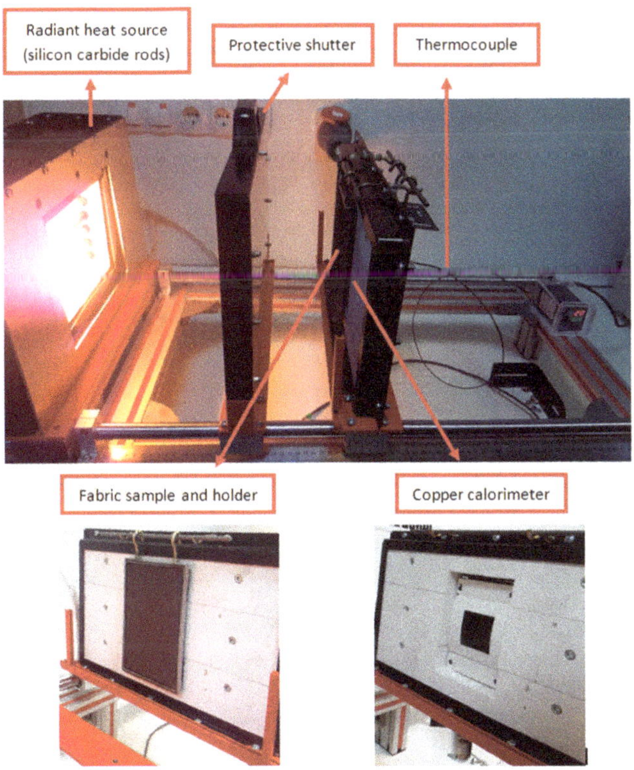

Figure 2. Radiant heat transmission measurement according to the EN ISO 6942 standard [18].

The second thermal measurement method was performed according to the JIS R 2618 standard [19] with a quick thermal conductivity meter (QTM-710) (Figure 3). This is the hot wire method (also called the probe method) and can determine the thermal conductivity of the samples by Equation (4):

$$\lambda = q \times ln(t_2 - t_1)/4\pi(T_2 - T_1) \tag{4}$$

where λ is the thermal conductivity of the sample (W/mK), q is the thermal unit of heater per time and length, t is time, and T is temperature. The measurement of samples for thermal conductivity was based on comparing the sample with three different reference plates (quartz, silicon, and polyethylene foam) in temperature rise when both were heated by the QTM-710 probe.

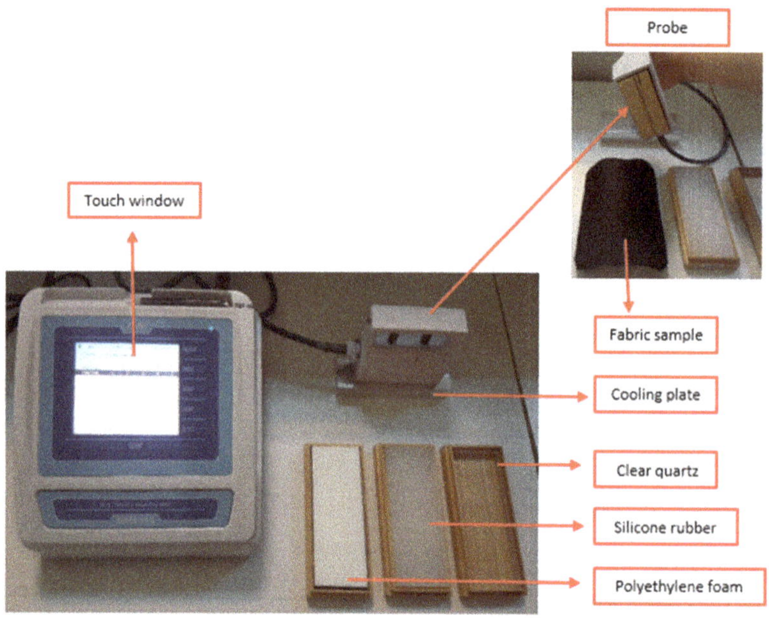

Figure 3. Thermal conductivity measurement according to the JIS R 2618 standard [19] with QTM-710.

The solar measurements were performed with a Shimadzu UV-3600 Plus spectrophotometer (Kyoto, Japan) with an integrating sphere at 280–2500 nm wavelength according to EN 14500:2008 [20]. Barium sulphate was chosen as a white reference according to the standard. The solar transmittance (T_S: 300–2500 nm), solar reflectance (R_S: 300–2500 nm), solar absorbance (A_S: 300–2500) and near-infrared reflectance (R_{NIR}: 800–2500 nm) values were calculated according to the EN 410 standard [21].

3. Results

3.1. Mass per Unit Area, Add-On, and Thickness Results

Figure 4a shows the increased mass per unit area and add-on values with increasing graphene concentration. These values also increased with increasing coating thicknesses both for references (R2.1 and R2.5) and graphene-coated fabrics at each concentration rate. The amounts of solid matter that were transferred increased with increasing concentration; thus, the mass per unit area values increased by 5.36%, 17.26% and 27.38% for GR50.5, GR100.5 and GR150.5, respectively, compared to R2.5 at 0.5 mm constant coating thickness. Add-on values increased by 11.66%, 37.58% and 59.61% for GR50.5, GR100.5 and GR150.5, respectively, compared to R2.5 at 0.5 mm constant coating thickness.

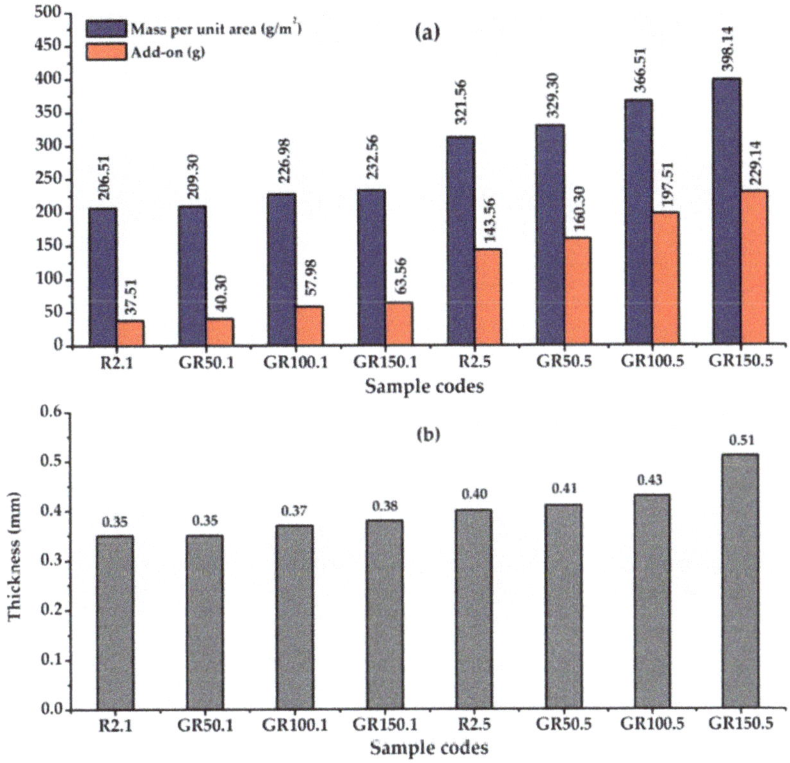

Figure 4. Mass per unit area and add-on results (**a**) and thickness results of reference and graphene-coated fabrics (**b**).

The thickness results of fabrics are given in Figure 4b. It was observed that the coating process generally increased the thickness values of the fabrics, as expected. The fabrics that coated 50 g/kg (GR50.1 and GR50.5) gave close results to the thickness values of relevant blind coatings (R2.1 and R2.5) at both 0.1 and 0.5 mm coating thicknesses. When the fabrics coated at maximum graphene concentration (GR150.1 and GR150.5) were evaluated, thickness values increased by 8.57% and 27.5% compared to R2.1 and R2.5, respectively.

3.2. Optical and SEM Image Results

Optical images (front and reverse sides) of the reference and graphene-coated fabrics are shown in Figure 5. It was observed that the coating process was performed successfully on the polyester base fabric. Additionally, the appearance of hills and valleys on fabric structure decreased with the increasing graphene concentration and coating thickness. The optical images of the reverse sides showed that the coating paste did not penetrate to the back side of the fabric. This was also desirable, because in tests measuring from the fabric surface, passing back the coating paste could negatively affect the results.

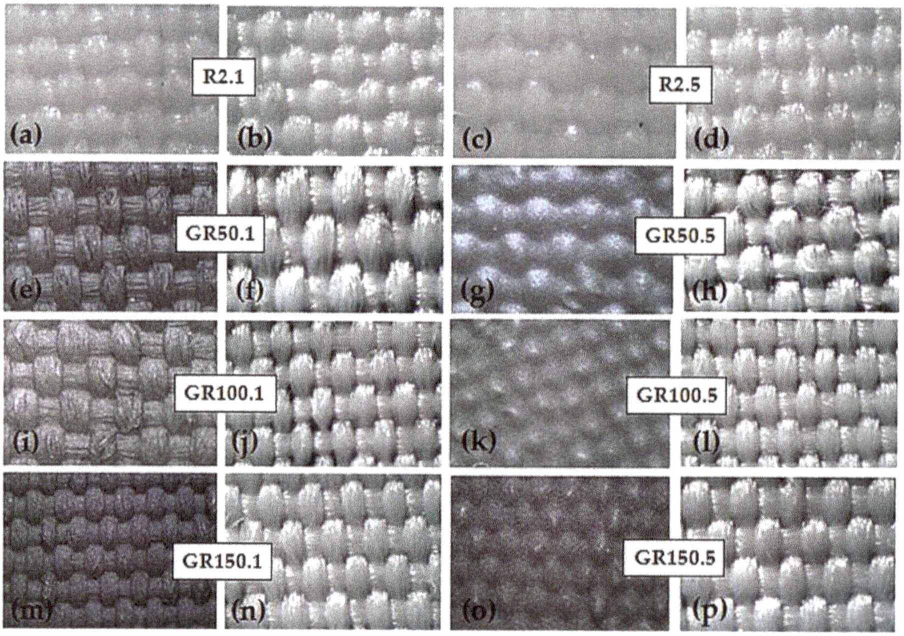

Figure 5. Front side images of fabrics (**a**,**c**,**e**,**g**,**i**,**k**,**m**,**o**) and reverse side images of fabrics (**b**,**d**,**f**,**h**,**j**,**l**,**n**,**p**).

SEM images of the reference fabrics are given in Figure 6 with 100× magnification rate. The effect of coating thickness was evident when the reference fabrics (R2.1 and R2.5) coated with stock paste were compared. The sectional views of graphene-coated fabrics with 150× and the surface views with 100× magnification rate on the top right corner are given in Figure 7. It was clearly seen that the coated surfaces were homogenous. The added paste amount increased with increasing graphene concentration and coating thickness, as expected. In Figure 7, when the surface images on the upper corner were compared between each other, it was seen that as the graphene concentration and coating thickness increased, the rate of filling the gaps between the peak and the valley at the intersection points of the fabric increased. Therefore, the surface coverage increased. It could be said that the increase in thickness played a more effective role in the difference in coverage compared to the increase in concentration. The highest coverage was achieved at the maximum concentration and thicker coating.

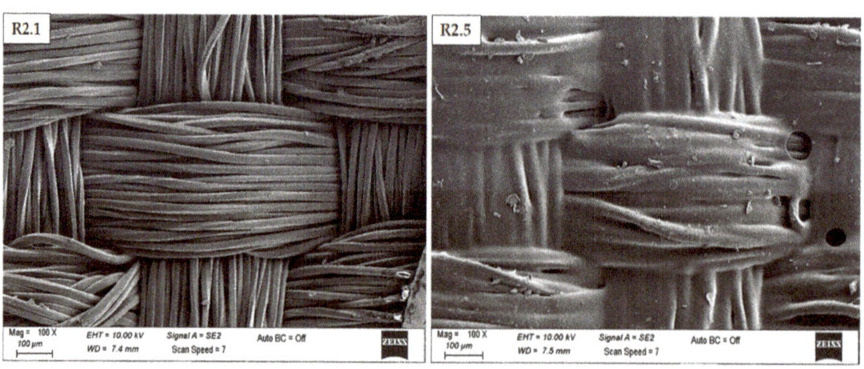

Figure 6. SEM images of reference fabrics for 0.1 and 0.5 mm coating thicknesses (Magnification: 100×).

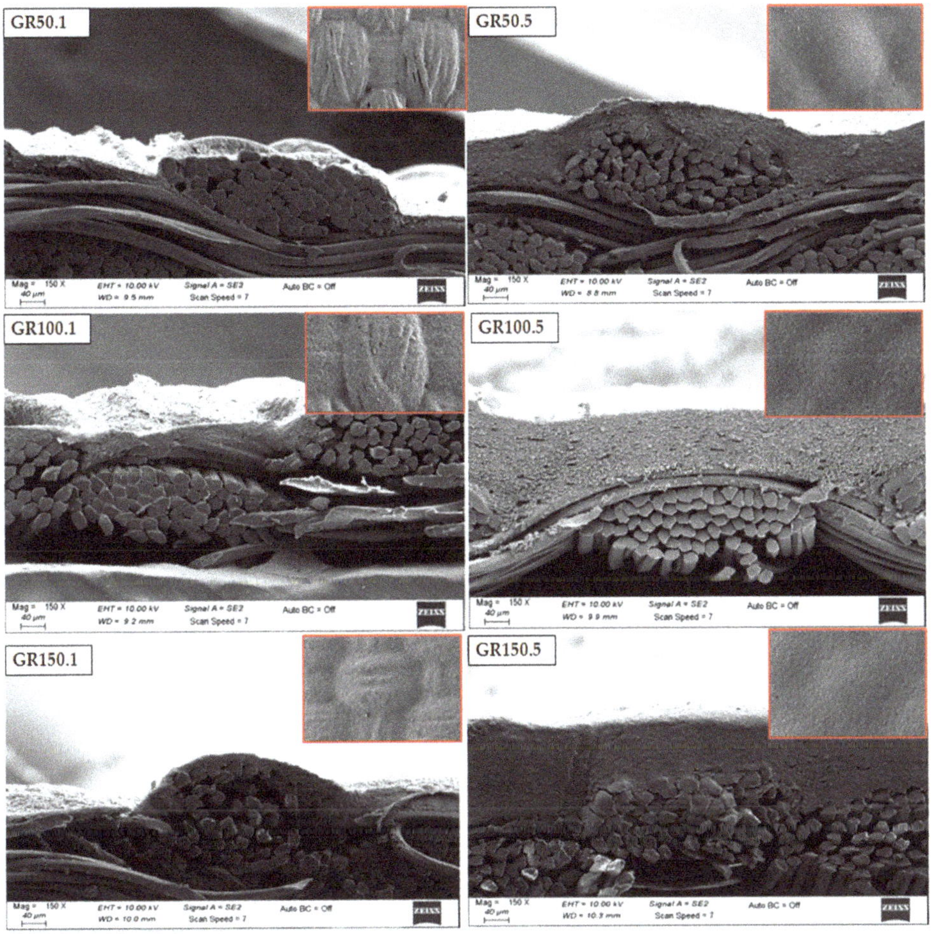

Figure 7. SEM images of graphene-coated fabrics (Mag: 150× for cross-section image).

3.3. Thermal Stability Results

The DSC test results for reference and graphene-coated samples are presented in Figure 8. There was no significant difference in the melting temperature values of the fabrics coated with graphene at different concentrations in different thicknesses compared to each other and the reference fabrics, similar to the study of Yang et al. [22]. All samples gave a peak at nearly 254 °C, showing the melting temperature of the polyester fabrics.

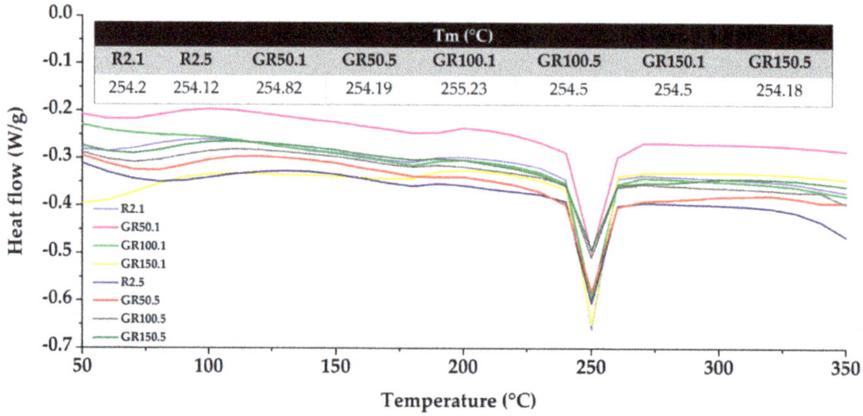

Figure 8. DSC curves of reference and graphene-coated fabrics.

TGA and DTG curves of samples with different coating thicknesses (0.1 and 0.5 mm) are given in Figure 9. The decomposition temperature range, DTG peak temperature, and total mass loss values are also presented in Table 6. The maximum decomposition rates are shown at the range of 350–450 °C. In this range, the mass losses were 76.03%, 55.03%, 54.76% and 52.4% for the R2.1, GR50.1, GR100.1 and GR150.1 samples, respectively, while the mass losses were 67.34%, 57.46%, 53.81% and 51.05% for the R2.5, GR50.5, GR100.5 and GR150.5 samples, respectively.

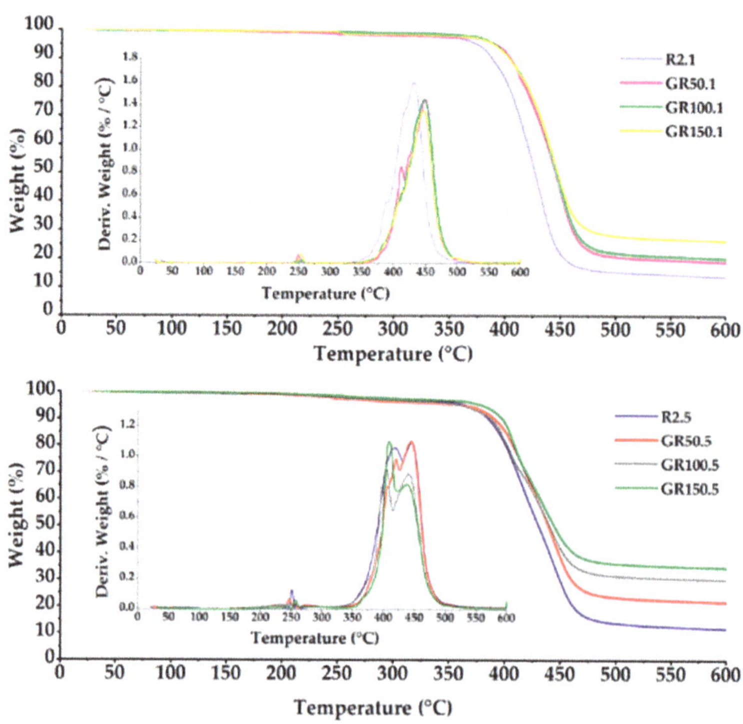

Figure 9. TGA and DTG curves of reference and graphene-coated fabrics for 0.1 and 0.5 mm coating thicknesses.

When the values were examined, it was seen that total mass loss decreased and thermal stability improved with the increase in both graphene concentration and coating thickness. The improvement in thermal stability was more significant, with increased coating thickness rather than increased concentration. This result was related to the total add-on amount (and, accordingly, the graphene content) and increased with thickness more than the concentration (Figure 4). The total mass loss at maximum concentration rates (GR150.1 and GR150.5) was reduced by 11.84% and 21.96%, respectively, compared to the related reference fabrics (R2.1 and R2.5). The TGA results conformed with some studies in the literature [13,23].

Table 6. TGA data of samples.

Sample	Temperature Range (°C)	DTG Peak Temperature (°C)	Mass Loss (%)	Total Mass Loss (%)
R2.1	200–300 350–450 450–600	246 432 -	0.85 76.03 7.85	84.73
GR50.1	200–300 350–450 450–600	249 448 -	1.04 55.03 23.91	79.98
GR100.1	200–300 350–450 450–600	254 446 -	0.43 54.76 23.71	78.90
GR150.1	200–300 350–450 450–600	253 445 -	1.02 52.40 19.47	72.89
R2.5	200–300 350–450 450–600	250 445 -	1.60 67.34 17.07	86.01
GR50.5	200–300 350–450 450–600	246 444 -	2.57 57.46 16.37	76.40
GR100.5	200–300 350–450 450–600	256 440 -	2.34 53.81 11.80	67.95
GR150.5	200–300 350–450 450–600	257 437 -	1.76 51.05 11.24	64.05

3.4. Thermal Conductivity and Spectrophotometry Results

Figure 10a shows the thermal conductivity coefficient (λ) values measured according to the hot wire method depending on the increasing graphene concentration and coating thickness. The fabrics coated at 0.1 mm coating thickness were examined; thermal conductivity values increased by 11%, 41%, and 87% for the GR50.1, GR100.1, and GR150.1 coded fabrics, respectively, compared to the R2.1 coded fabric. When the results were examined for 0.5 mm coatings, thermal conductivity values increased by 38%, 96%, and 262% for the GR50.5, GR100.5 and GR150.5 coded fabrics, respectively, compared to the R2.5 coded fabric. These results were consistent with other studies [14,15,24], showing that an increase in graphene concentration increases thermal conductivity.

The thickness is an important parameter that affects thermal conductivity. In the literature, some studies showed that the thermal conductivity changed directly proportionally with the coating thickness [25–27]. When the effect of coating thickness for the same concentrations was evaluated, conductivity values increased by 47%, 65% and 129% for 50, 100 and 150 g/kg rates, respectively, when the coating thickness increased from 0.1 to 0.5 mm. Therefore, these results are in-line with the previous studies.

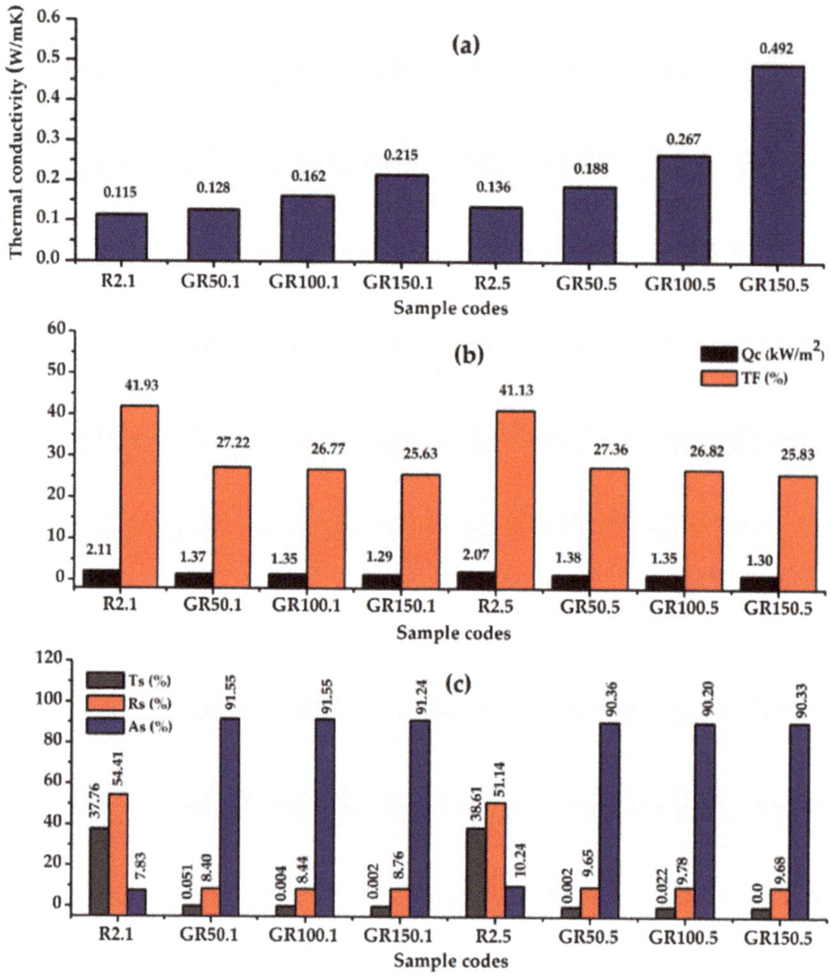

Figure 10. Thermal results according to the JIS R 2618 standard [19] (**a**); thermal results according to the EN ISO 6942 standard [18] (**b**); and spectrophotometry results of the reference and graphene-coated fabrics [20,21] (**c**).

Figure 10b shows the transmitted heat flux density (Q_C) and heat transmission factor (*TF*) values depending on the increasing graphene concentration and coating thickness.

When the graphene-coated fabrics were compared with each other, it was observed that the difference in coating thickness did not cause a significant change in the Q_C results. For a constant thickness value, the Q_C values decreased slightly as the graphene concentration increased. As can be seen from Equation (3), Q_C and *TF* values are directly proportional to each other. Therefore, *TF* values also showed parallel changes with Q_C.

It is known that graphene has a high thermal conductivity, and our results according to the JIS R 2618 standard were also consistent with this. However, according to the radiant heat transmission test results, there was no improvement in heat transmission with the increasing graphene concentration and coating thickness; on the contrary, the Q_C and *TF* values of the graphene-coated samples decreased compared to the reference fabrics. To interpret the results more comprehensively, considering the energy in the radiant heat test it was necessary to examine the reflectance behavior of the samples in the near-infrared region. For this reason, the solar absorbance, reflectance, and transmittance values of the samples were also measured.

Figure 10c shows the solar transmittance (T_S), reflectance (R_S) and absorbance (A_S) results of the graphene-coated and reference fabrics. Graphene concentration and coating thickness did not make a significant difference in transmittance values. Even with the lowest concentration and coating thickness, the solar transmittance values approached zero. The T_S values of the samples at the maximum graphene concentration (GR150.1 and GR150.5) were almost zero and decreased by 37.76% and 38.61% compared to reference fabrics at 0.1 and 0.5 mm coating thickness, respectively. Similarly, when the graphene-coated fabrics were evaluated within themselves, there was no significant difference in reflectance results. With increasing graphene concentration compared to the R2.1 and R2.5, the R_S values decreased by 46.01%, 45.97%, 45.65% and 41.49%, 41.36%, 41.46% for 0.1 mm and 0.5 mm coating thickness, respectively. It is known that carbon nanomaterials (graphene, carbon black, etc.) exhibit good optical absorption properties due to their dark colour [28]. In parallel with this fact, the solar absorbance (A_s) values of graphene-coated samples dramatically increased compared to the reference fabrics, even at the lowest graphene concentrations. The maximum increase rates were 83.72% and 80.12% at 0.1 and 0.5 mm thickness, respectively. All graphene-coated samples had close A_S values. It was thought that because the colour reached a certain saturation with the lowest graphene concentrations, there was no significant increase in the absorbance values after 50 g/kg with increasing thickness and concentration.

The rays in the near-infrared region play an effective role in the radiant heat transmission test results mentioned above; therefore, the near-infrared reflectance (R_{NIR}) behaviour was investigated separately from the R_S. R_{NIR} values tended to increase slightly with increasing concentration at a constant thickness, as shown in Figure 11. The increase was more apparent with increasing thickness at a constant concentration. While graphene powder had a 12.39% R_{NIR} value, the closest sample to it was the GR150.5 sample with 11.23% R_{NIR}.

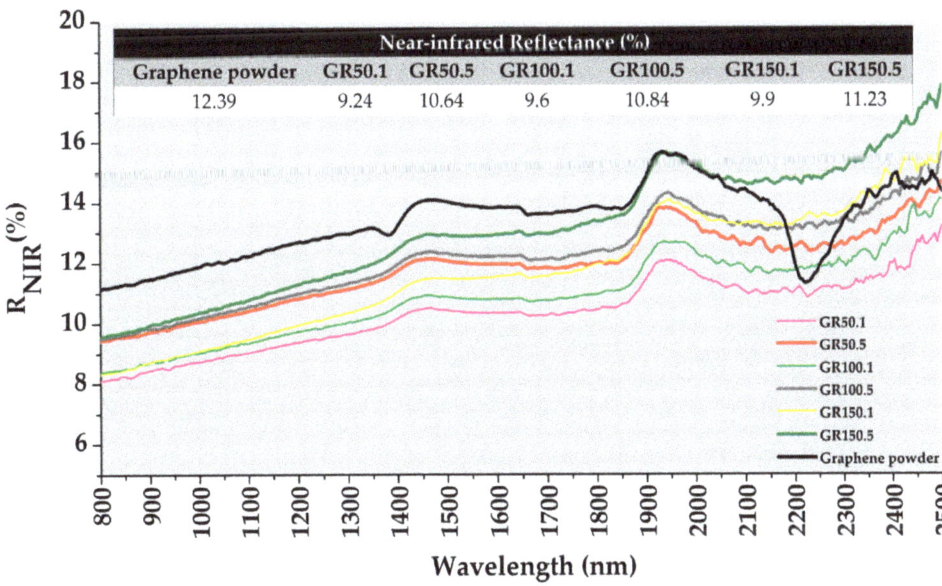

Figure 11. Near-infrared reflection (R_{NIR}) results of graphene powder and coated fabrics.

When the radiant heat transmission results (Q_C, TF) given in Figure 10b and solar results were evaluated together, the following conclusions have been reached. Firstly, it was thought that the graphene-coated samples might have lower heat flux density and heat transmission factor values compared to the references due to their relatively high absorbance values. Secondly, it was concluded that the TF values, which decreased slightly with the increasing graphene concentration at the constant thickness, might be caused by the reflectance values showing small rises with increasing concentration in the near-infrared region.

4. Conclusions

The graphene nano platelet powder was successfully applied to the polyester woven fabrics by a knife-over-roll coating process. The effects of graphene concentration and coating thickness on the thermal stability, heat transmission behaviour, and solar properties of samples were examined. The structure and morphology of the coated samples were characterized by optical and SEM analysis. It was demonstrated that homogenous coating surfaces were obtained. While there was no significant change in the melting temperature according to DSC analysis, TGA results showed that graphene coating improved thermal stability.

In summary, the thermal conductivity results performed according to two different standards (JIS R 2618 and EN ISO 6942); in the contact heat transfer, extremely high rises in thermal conductivity values were observed in parallel with the increase in coating thickness and concentration. While graphene-coated fabrics showed lower radiant heat transmission rate compared to reference fabrics due to their high solar absorbance value, the heat transmission values of graphene-coated fabrics were close to each other due to slightly increased R_{NIR} values.

As well as widely known application areas, graphene, which has superior mechanical, electronic, thermal, and optical properties, can be used in the functional textile coating and it is promising in contributing to multidisciplinary studies and open for development.

Author Contributions: Conceptualization, G.M., R.C., M.K. and Y.U.; methodology, G.M. and R.C.; investigation, G.M., R.C., M.K. and Y.U.; writing—original draft preparation, G.M. and R.C.; writing—review and editing, G.M., R.C., M.K. and Y.U.; supervision, M.K. and Y.U. All authors have read and agreed to the published version of the manuscript.

Funding: This research received no external funding.

Institutional Review Board Statement: Not applicable.

Informed Consent Statement: Not applicable.

Data Availability Statement: The data presented in this study are available on request from the corresponding author.

Conflicts of Interest: The authors declare no conflict of interest.

References

1. Giessmann, A. *Coating Substrates and Textiles: A Practical Guide to Coating and Laminating Technologies*; Springer Science & Business Media: Heidelberg, Germany, 2012; p. 101.
2. Singha, K. A review on coating & lamination in textiles: Processes and applications. *Am. J. Polym. Sci.* **2012**, *2*, 39–49.
3. Qian, L.; Hinestroza, J.P. Application of nanotechnology for high performance textiles. *J. Text. Appar. Technol. Manage.* **2004**, *4*, 1–7.
4. Gong, J.R. *Graphene-Synthesis Characterization, Properties and Applications*; InTech.: Rijeka, Croatia, 2011.
5. Introduction to the Physical Properties of Graphene. Available online: http://web.physics.ucsb.edu/~{}phys123B/w2015/pdf_CoursGraphene2008.pdf (accessed on 26 November 2020).
6. Radadiya, T.M. A properties of graphene. *Eur. J. Mater. Sci.* **2015**, *2*, 6–18.
7. Shahil, K.M.; Balandin, A.A. Thermal properties of graphene and multilayer graphene: Applications in thermal interface materials. *Solid. State. Commun.* **2012**, *152*, 1331–1340. [CrossRef]
8. Summary of Graphene (and Related Compounds) Chemical and Physical Properties. Available online: https://assets.publishing.service.gov.uk/government/uploads/system/uploads/attachment_data/file/627145/Summary_of_Graphene__and_Related_Compounds__Chemical_and_Physical_Properties.pdf (accessed on 5 July 2020).
9. Hu, X.; Tian, M.; Qu, L.; Zhu, S.; Han, G. Multifunctional cotton fabrics with graphene/polyurethane coatings with far-infrared emission, electrical conductivity, and ultraviolet-blocking properties. *Carbon* **2015**, *95*, 625–633. [CrossRef]

0. Lin, Y.; Jia, Y.; Alva, G.; Fang, G. Review on thermal conductivity enhancement, thermal properties and applications of phase change materials in thermal energy storage. *Renew. Sust. Energ. Rev.* **2018**, *82*, 2730–2742. [CrossRef]
1. Sang, M.; Shin, J.; Kim, K.; Yu, K.J. Electronic and thermal properties of graphene and recent advances in graphene based electronics applications. *Nanomaterials* **2019**, *9*, 374. [CrossRef] [PubMed]
2. Balandin, A.A. Thermal properties of graphene and nanostructured carbon materials. *Nat. Mater.* **2011**, *10*, 569–581. [CrossRef] [PubMed]
3. Gan, L.; Shang, S.; Yuen, C.W.M.; Jiang, S.X. Graphene nanoribbon coated flexible and conductive cotton fabric. *Compos. Sci. Technol.* **2015**, *117*, 208–214. [CrossRef]
4. Abbas, A.; Zhao, Y.; Zhou, J.; Wang, X.; Lin, T. Improving thermal conductivity of cotton fabrics using composite coatings containing graphene, multiwall carbon nanotube or boron nitride fine particles. *Fiber. Polym.* **2013**, *14*, 1641–1649. [CrossRef]
5. Gunasekera, U.; Perera, N.; Perera, S.; Hareendra, Y.; Somaweera, L.; De Silva, N.; Tissera, N.; Wijesinghe, R. Modification of Thermal Conductivity of Cotton Fabric Using Graphene. In Proceedings of the Moratuwa Engineering Research Conference, University of Moratuwa, Moratuwa, Sri Lanka, 7–8 April 2015.
6. *TS 7128 EN ISO 5084-Textiles- Determination of Thickness of Textiles and Textile Products*; Turkish Standard Institution: Ankara, Turkey, 1998.
7. *TS 251- Determination of Mass per Unit Length and Mass per Unit Area of Woven Fabrics*; Turkish Standard Institution: Ankara, Turkey, 1991.
8. *EN ISO 6942- Protective Clothing—Protection Against Heat and Fire—Method of Test: Evaluation of Materials and Material Assemblies When Exposed to a Source of Radiant Heat*; International Organization for Standardization: Geneva, Switzerland, 2002.
9. *JIS R 2618-Testing Method for Thermal Conductivity of Insulating Fire Bricks by Hot Wire*; Japanese Standards Association: Tokyo, Japan, 1992.
0. *EN 14500:2008-Blinds and Shutters- Thermal and Visual Comfort-Test and Calculation Methods*; European Committee for Standardization: Brussels, Belgium, 2008.
1. *EN 410-Glass in Building- Determination of Luminous and Solar Characteristics of Glazing*; European Committee for Standardization: Brussels, Belgium, 2011.
2. Yang, W.; Zhang, L.; Guo, Y.; Jiang, Z.; He, F.; Xie, C.; Fan, J.; Wu, J.; Zhang, K. Novel segregated-structure phase change materials composed of paraffin@ graphene microencapsules with high latent heat and thermal conductivity. *J. Mater. Sci.* **2018**, *53*, 2566–2575. [CrossRef]
3. Moharram, M.A.; Ereiba, K.M.T.; El Hotaby, W.; Bakr, A.M. Thermal degradation studies of graphene oxide polymer composite. *Middle East. J. Appl. Sci.* **2015**, *5*, 23–30.
4. Manasoglu, G.; Celen, R.; Kanik, M.; Ulcay, Y. Electrical resistivity and thermal conductivity properties of graphene-coated woven fabrics. *J. Appl. Polym. Sci.* **2019**, *136*, 48024. [CrossRef]
5. Zhao, J.; Du, F.; Cui, W.; Zhu, P.; Zhou, X.; Xie, X. Effect of silica coating thickness on the thermal conductivity of polyurethane/SiO2 coated multiwalled carbon nanotube composites. *Compos. Part A Appl. Sci. Manuf.* **2014**, *58*, 1–6. [CrossRef]
6. Rätzer-Scheibe, H.J.; Schulz, U.; Krell, T. The effect of coating thickness on the thermal conductivity of EB-PVD PYSZ thermal barrier coatings. *Surf. Coat. Technol.* **2006**, *200*, 5636–5644. [CrossRef]
7. Griffin, A.J., Jr.; Brotzen, F.R.; Loos, P.J. Effect of thickness on the transverse thermal conductivity of thin dielectric films. *J. Appl. Phys.* **1994**, *75*, 3761–3764. [CrossRef]
8. Chen, L.; Xu, C.; Liu, J.; Fang, X.; Zhang, Z. Optical absorption property and photo-thermal conversion performance of graphene oxide/water nanofluids with excellent dispersion stability. *Sol. Energy* **2017**, *148*, 17–24. [CrossRef]

coatings

MDPI

Article

Development of Advanced Textile Finishes Using Nano-Emulsions from Herbal Extracts for Organic Cotton Fabrics

Prabhuraj D. Venkatraman [1,*], Usha Sayed [2], Sneha Parte [2] and Swati Korgaonkar [2]

1 Manchester Fashion Institute, Manchester Metropolitan University, Cavendish Street, Manchester M15 6BG, UK

2 Department of Fibres and Textile Chemistry, Institute of Chemical Technology (ICT), Nathalal Parekh Marg, Matunga, Mumbai 400 019, India; u.sayed@ictmumbai.edu.in (U.S.); snhprt37@gmail.com (S.P.); swatikorgaonkar25@gmail.com (S.K.)

* Correspondence: p.venkatraman@mmu.ac.uk

Abstract: The development of textile finishing with improved functional properties has been a growing interest among industry and scientists worldwide. The recent global pandemic also enhanced the awareness amongst many toward improved hygiene and the use of antimicrobial textiles. Generally, natural herbal components are known to possess antimicrobial properties which are green and eco-friendly. This research reports a novel and innovative method of developing and optimising nano-emulsions using two combinations of herbal extracts produced from *Moringa oleifera*, curry leaf, coconut oil (nano-emulsion 1) and other using *Aegle marmelos* with curry leaf and coconut oil (nano-emulsion 2). Nano-emulsions were optimised for their pH, thermal stability, and particle size, and percentage add-on. Organic cotton fabrics (20 and 60 gsm) were finished with nano-emulsions using continuous and batch processes and characterised for their surface morphology using scanning electron microscopy, energy dispersive X-ray (EDX) analysis and Fourier transform infrared spectroscopy (FTIR) analysis. The finished fabrics were evaluated for their Whiteness Index, assessed for antimicrobial resistance against Gram-positive (*Staphylococcus aureus*) and Gram-negative bacteria (*Escherichia coli*) using AATCC 100 and 147 methods. In addition, fabrics were assessed for their antifungal efficacy (AATCC 30), tensile strength and air permeability. Results suggested that finished organic fabrics with nano-emulsions had antimicrobial resistance, antifungal, wash fastness after 20 washing cycles, and sufficient strength. This novel finishing method suggests that organic cotton fabrics treated with nano-emulsions can be used as a durable antimicrobial textile for healthcare and hygienic textiles.

Keywords: *Moringa oleifera*; *Aegle marmelos*; organic cotton; nano-emulsion; antimicrobial resistance; antifungal efficacy; wash fastness

Citation: Venkatraman, P.D.; Sayed, U.; Parte, S.; Korgaonkar, S. Development of Advanced Textile Finishes Using Nano-Emulsions from Herbal Extracts for Organic Cotton Fabrics. *Coatings* **2021**, *11*, 939. https://doi.org/10.3390/coatings11080939

Academic Editor: Chi-wai Kan

Received: 16 June 2021
Accepted: 2 August 2021
Published: 5 August 2021

1. Introduction

Cotton based textile fabrics retain sufficient moisture [1], offer a large surface area, can absorb moisture from the environment and human body and maintain body temperature. These properties of cotton fabrics serve as an ideal environment for the growth of fungi and microbes. Studies have shown the survival of several Gram-positive bacteria (*Staphylococcus aureus*, *Enterococcus faecalis*) on standard hospital fabrics made of 100% cotton clothing, 100% cotton terry towels, 60%/40% cotton/polyester-scrub suits and lab coats, and 100% polyester drapes. Swatches were inoculated with micro-organisms and examined for growth over 48 h. Most bacterial growth survived at least a day, and some survived more than 90 days [2]. High survival of bacteria through textiles leads to the spread of infections requiring proper hygiene, procedures to control infections, and disinfection of contact surfaces and textiles [3]. This also shows that textiles made of natural fibres (cotton) are more susceptible to microbial growth, especially when they contact the body and moist objects leading to transmission of infections [3]. In order to counteract the microbial attack, textiles intended for medical and hygiene applications are treated with a range of antimicrobial finishes obtained from chemicals, including quaternary ammonium compounds, triclosan, metallic salts (TiO_2, ZnO), chitosan, phenolic compounds, oxidising agents, and poly (hexamethylene biguanide) [4].

Textiles (cotton, silk, and bamboo) were treated with natural extracts such as resveratrol, a natural polyphenol, obtained from *Polygonum cuspidatum* (Japanese knotweed), also widely found in fruits, such as grapes, red wine, mulberries, and other fruits and were reported to possess pharmacological activities—Antibacterial, antifungal, and anti-inflammatory properties [5]. Bio-functional textiles can be defined as those textiles treated with biocompatible materials with drug delivery systems, widely preferred in medical applications [6]. In addition, bio-functional textiles can absorb from the skin or release therapeutic or cosmetic compounds [7]. Textiles are finished using a variety of methods and techniques, however in addition to the traditional finishing techniques, padding or continuous process (involves dipping the fabric in the bath and squeezing between the rollers) and exhaust or batch process (involves immersion of fabrics in the liquor bath and rotating in the bath for a given period), layer-by-layer deposition method has been recently developed. This involves dipping the textiles in an electrically charged electrolyte where a monolayer of the film is deposited [8]. Other innovative methods of developing bio-functional textiles include using a confined impinging jet mixer (CIJM) to produce nanoparticles of menthol loaded poly-e-caprolactone [9]. Furthermore, ultrasound irradiation assisted water/oil/water microemulsion method of producing nano-capsules (60–80 nm) of chamomile extracts on cotton fabrics using UV curing were also reported to have good antimicrobial properties [10]. Recently, nanoparticles were developed using the flash nanoprecipitation technique (FNP) to produce bio-functional cotton/micro-modal textiles with caffeine for transdermal applications [11].

The FNP method has also been demonstrated to entrap caffeine in poly-ε-caprolactone with control release [12]. Various nanoparticles were deposited on polyester/cotton or cotton fabrics with gold, silver, ZnO, or TiO_2 particles showing good antibacterial and antifungal activity [13]. A decrease in particle size shows better antibacterial results due to the high surface area to volume ratio and increases its wash durability, and the adhesion of nanoparticles onto the textiles can be affected using surfactant [13]. Several methods of surface modification of textiles with antimicrobial properties were also highlighted recently, including nanoparticle technology, micro and nano-capsules, vapour deposition, sol–gel, and encapsulation [14].

Antimicrobial chemicals or material can inhibit bacterial cell growth, called the biostatic effect, or kill micro-organisms known as the biocidal effect [15,16]. Most antimicrobial agents used in textiles are biocides [17]. Biostatic inhibits micro-organisms' growth, prevents odour control, and preserve textiles leaving live bacteria on the surface. Although the infection rate can be reduced, it leaves the residual micro-organisms that could cause transmission of infection. Biocidal kill the micro-organism in a short contact time, provide complete disinfection of diseases on the surface, and prevent transmission of infections in a hospital setting [18]. The resistance level also depends on the dosage of the antimicrobial agent, the affinity of the antibacterial agent finishing on textiles and its end-use.

Many synthetic chemicals are toxic and may not decompose quickly, affecting aquatic organisms when washed into streams. When released from textiles into waste waterways, synthetic antimicrobial agents can affect the aquatic environment when not recovered or recycled. The biodegradability of QACs (quaternary ammonium compounds) is poor [19]. Degradation of triclosan is also a concern and can be toxic, and research also highlighted environmental and health problems of various antimicrobials (triclosan, quaternary ammonium compounds, and copper) and risk associated with triclosan [20]. The most silver-based and mineral-based product cannot be biodegradable [21], as it is a mineral substance. It was reported that many synthetic chemicals could ideally be replaced with natural or plant-based compounds, which do not have any adverse effects [4]. The plant-based herbal oils for environmentally-friendly antimicrobial efficacy have been highlighted [4,22]. Plant-based substances provide an efficient antimicrobial resistance. They are safe, widely available, non-toxic, and environmentally friendly [23,24]. In addition, the application of enzymes in textile wet processing had been reported to have many advantages, such as reduced use of chemicals, water, and energy, resulting in eco-friendly products [25]. The use of biopolymers (such as chitosan, chitosan nanoparticles, cyclodextrins) for finishing cellulose fabrics with functional properties (antibacterial) have also been reported [26].

Ideally, antimicrobial textiles should inhibit the growth of a broad range of bacteria and fungi, exhibit low toxicity, non-allergic, meet standards for compatibility—cytotoxicity, irritation, and sensitisation. In addition, the finish should be durable during washing, possess strength, appearance, and be compatible with textile chemical processing—dyeing, and not harmful to the environment [15,27]. Textile treated biocides either protect the textile or user from microbial growth. Due to recent environmental awareness, several natural bioactive agents, such as chitosan, neem, aloe vera, tea tree, prickly chaff flower, eucalyptus oil, clove oil, curcumin, hiba oil, and natural dyeing materials were of focus, and studies have discussed the development of eco-friendly and antimicrobial textiles [28]. Various natural dyes [29] were also evaluated for the antimicrobial finish of cotton fabrics against Gram-positive and Gram-negative bacteria. Among many natural dyes, *Q. infectoria* (Aleppo Oak) natural dye extract showed good resistance (70–90%) to most microbes when combined with alum or copper mordants against 45–60% inhibition of bacterial growth without mordants. In addition, tannin, a water-soluble polyphenol found in many plant and tree species (bark, leaf, root), also possess antimicrobial activity against a wide range of bacteria and fungi [30]. Evaluation of total phenolic content and flavonoids content using leaf extracts (eucalyptus and lemon grass) with methanol, ethanol, chloroform, and distilled water extract indicated the potential of antimicrobial application of textiles [31]. There were also reports on the use of curry leaf and ginger oil showing good antibacterial and antifungal resistance on plain-woven cotton fabric [32].

Moringa oleifera, which belongs to the Moringaceae family, commonly called drumstick tree or horseradish tree, grows in tropical and subtropical regions widely found in India. Moringa leaves and drumsticks are widely consumed as a healthy food by Asians [33] Leaves of *M. oleifera* is rich in calcium, potassium, zinc, magnesium, iron, and copper [34]. Apart from its rich nutritive potential, the leaves of *M. oleifera* possess flavonoids that give its antioxidant properties. It is used to treat many diseases, including asthma, flu, diarrhoea, headaches, skin diseases, eye and ear infections [34]. A recent study highlighted the antimicrobial resistance to *Bacillus subtilis*, *Staphylococcus aureus* and *Vibrio cholera* using extracts prepared from moringa seeds [35]. Rockwood et al. (2013) reported the antimicrobial efficacy of *M. oleifera* seeds and leaves using different solvents—deionised water, inorganic ethanol and organic ethyl acetate on 14 micro-organisms [36]. They reported that leaf and seed extracts prepared with ethanol and ethyl acetate showed no inhibition. However, leaf extracts with deionised water showed inhibition against *Bacillus sphaericus*, and seed extracts inhibited *Bacillus sphaericus*, *Mycobacterium smegmatis*, *S.aureus* and *Alcaligenes faecalis*. They concluded that Moringa seed extracts were effective than leaf extracts inhibiting bacterial growth. More recently, the use of fresh leaves of *Moringa oleifera* prepared by water and methanol extracts showed varied levels of antibacterial activity against *P. aeruginosa*, *Klebsiella pp.*, *E. coli* and *Staphylococcus spp.* isolates [37].

Similarly, *Aegle marmelos* (also commonly called Bael), widely grown in India—outer Himalayas, Shivaliks, South Indian plateau, has shown potential antimicrobial properties. Its leaves, fruits, stems and roots are used for treating various ailments [38]. Several compounds have been noted with Bael, but eugenol [39] and cuminaldehyde [40] are responsible for providing antibacterial properties. Cuminaldehyde is a class of benzaldehydes substituted by isopropyl alcohol at position 4 [41] (see Figure 1). Leaf extracts of Bael have shown antibacterial resistance against *Escherichia coli* [42,43]. Oil extracts of Bael also have shown resistance against *Pseudomonas salacearum*, *Xanthomonas vesicatoria*, *Escherichia coli*, and *Aeromonas sp.* [44,45]. Therefore, it can be inferred that both *Moringa oleifera* and *Aegle marmelos* possess antimicrobial resistance. However, neither of these herbal extracts have been reported for their potential on cotton-based textiles, particularly for their durability and other physical and comfort properties. Based on the above findings, this project focuses on the herbal-based finishing of cotton-based fabrics with antibacterial and antifungal properties, which can be widely produced and used in the community. The application of fabrics with nano-emulsions by the continuous or batch process is available in the local community (urban and rural parts of India). Nano-emulsion shown in this study is easy to prepare (using a simple homogeniser—a high-speed stirrer). Other new methods, comparatively, require more considerable investment resulting in high-end fabrics. Therefore, the padded or exhaust finished fabrics allow the adsorption of sufficient nano-emulsion on the cotton fabrics, producing durable finishing with antibacterial properties. This study reports for the first time the development of herbal nano-emulsions using *Moringa oleifera* and *Aegle marmelos* blended with curry leaves and coconut oil. Organic cotton fabrics (varying fabric density) were finished with the above nano-emulsion using continuous (padding) and batch methods (exhaust), and their antimicrobial resistance before wash, after 10 and 20 wash cycles, are presented. The particle size analysis, the pH, the thermal stability, and the percentage add-on of the nano-emulsions were analysed to determine the optimum processing conditions. The main novelty in this project is in the development of a unique combination of herbal nano-emulsions (using *Moringa oleifera*, *Aegle marmelos*, curry leaves and pure coconut oil), its application on organic cotton fabrics and the antibacterial and antifungal resistance of finished fabrics to a wide range of Gram-positive, Gram-negative bacteria and fungi.

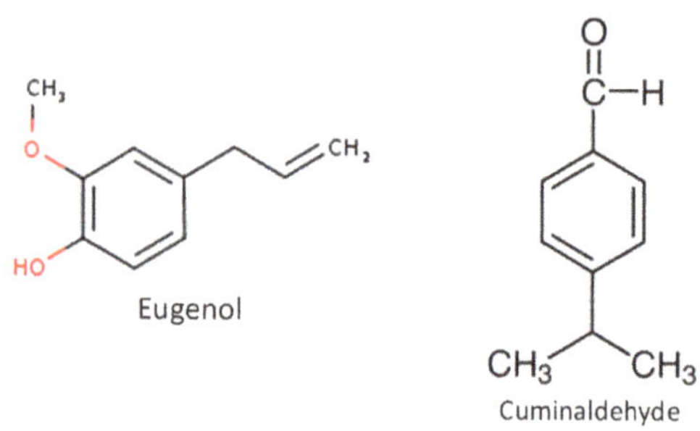

Figure 1. Compounds of *Aegle marmelos*.

2. Materials and Test Methods

2.1. Materials

Organic cotton fabrics—20 g/m^2 (plain weave) and 60 g/m^2 (twill weave) certified by GOTS (Global Organic Test Standards) was supplied by Test fab India (Vapi, Gujarat, India). These lightweight fabrics were selected with the scope of using the finished fabrics as durable wipes. The leafy herbs—*Moringa oleifera*, *Aegle marmelos*, curry leaves (*Murraya koengii*), and odourless pure coconut oil were purchased from Matunga, Mumbai. The surfactant polysorbate monobate 80 (a natural vegetable emulsifier) was purchased from LOBA Chemicals, Mumbai, India, while ethanol was purchased from Himedia Chemicals, Mumbai, India. Polysorbate 80 is a non-ionic surfactant, a viscous liquid which has a yellow to amber colour and is approved by FDA (The Food and Drug Administration, US) [46]. It is widely used in cosmetic, food, and pharmaceuticals industries [47] (Figure 2).

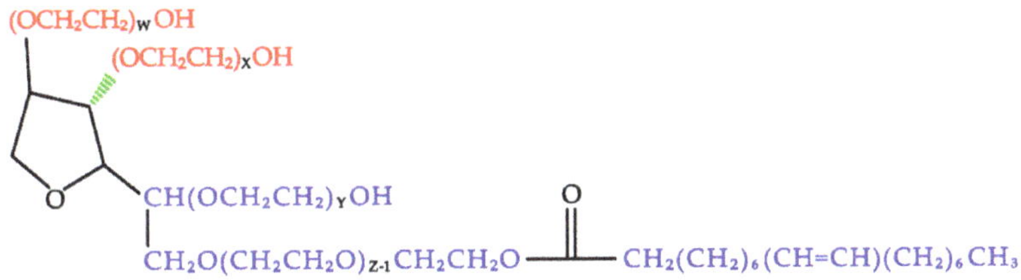

Where W + X + Y + Z = 20 represents one monomer

Figure 2. Polysorbate 80 (glycol).

The selection of herbal combinations (*Moringa oleifera*, *Aegle marmelos*, curry leaf, coconut oil) was based on an extensive survey of herbal constituents, and the proportions of each constituent were based on several trials. *Aegle marmelos* and *Moringa oleifera*, curry leaves and coconut oil are consumed as food substances in routine Asian cuisine [34,48–50], and it is anticipated that these nano-emulsions will not cause any adverse effects. Therefore, the fresh leaves of *Moringa oleifera* and *Aegle marmelos* were used to prepare different herbal extracts and to this extract, curry leaves and coconut oil were added to produce the two combinations—nano-emulsion *1* (*Moringa oleifera*, curry leaf and coconut oil) and nano-emulsion *2* (*Aegle marmelos*, curry leaf and coconut oil).

2.2. Methodology of Herbal Extraction

i. Extraction of oil: Both the herbs (*Moringa oleifera* and *Aegle marmelos*) were washed thoroughly with the distilled water and dried in the oven at 105 °C for one hour to remove all the dirt and impurities.
ii. Steam Distillation: The dried herb of 10 gm of *Moringa oleifera*, 5 gm of curry leaves and 100 mL coconut oil have been boiled by heating these constituents using steam supplied from a steam generator. The heat applied determines how effectively the plant material structure breaks down and bursts and releases the aromatic components of essential oils. Thus, the steam distillation extraction technique increases the isolated essential oil yields and reduces wastewater produced during the extraction process.
iii. Solvent extraction: The mixture is further used for the extraction of oil through the solvent extraction technique. The solvent used for extraction is 99% pure ethanol. The dried herbs are kept in the thimble of the Soxhlet extractor, and ethanol solution has been added. The extracted solution has been collected in the collector. The collected oil is then filtered, and once the solvent is evaporated, leaving the oil in the pot as residue.
iv. The extract yield has been calculated by:

$$\text{Oil yield} = \frac{\text{Amount of extracted oil (g)}}{\text{Amount of dry herbs and oil (g)}} \times 100\%$$

The second set of oil consisting of *Aegle marmelos*, curry leaves, and coconut oil mixture has been extracted in the same way. The extracted oil has been stored in a glass bottle until further analysis.

v. Preparation of nano-emulsion

The herbal oil and the surfactant were prepared in the following ratios 1:0.5, 1:1, 1:1.5, 1:2, 1:2.5. Nano-emulsions of these five ratios were prepared using a high-speed homogeniser (Tool-Tech), which mixes at 1000 to 5000 rpm. The homogenisation was carried out for one hour for each nano-emulsion and for all the ratios to obtain a stable and uniform nano-emulsion. The 10-, 20-, and 30-g-per-litre concentrations of nano-emulsions were characterised using particle size measurements. The emulsions have also been evaluated for their pH to determine their stability. As the level of surfactant increased, the particle size decreased, enabling it to penetrate the fabric more easily. Thermal stability was evaluated by maintaining the emulsions at varying temperatures, observing their homogenous nature, and recording any oil separation from the constituents. Visual methods (using a separating funnel) were used to observe the oil separation. The emulsion breaks and separates into immiscible compounds on reaching a particular temperature (>60 °C). The stability of nano-emulsion 1 (combination of *Moringa oleifera, curry leaves, and coconut oil*) and nano-emulsion 2 (combination of *Aegle marmelos, curry leaves, and coconut oil*) was assessed and discussed later in Section 3.3. In this manner, emulsions have been characterised and optimised for the application of cotton fabrics. In the case of 1:1 ratio, 100 mL of distilled water, one ml of oil mixture (3 parts of *Moringa oleifera*, two parts of curry leaves, and one part *coconut oil*), and one ml of polysorbate 80 was used. This combination is termed as nano-emulsion 1. Similarly, for 1:1 ratio, *Aegle marmelos* was used in place of *Moringa oleifera* to produce nano-emulsion 2.

A particle size analyser (Shimadzu SALD- 7500 nano, Kyoto, Japan) was used for determining the particle size of the nano-emulsion of the herbal extracts. The Whiteness Index of the samples was measured using Spectra scan 5100+, Computer colour matching system (Rayscan, Liverpool, Australia). The test method uses the amount of light reflected from the fabric surface at each wavelength to give the Whiteness Index. A Meta–Lab MSI-17B (Meta-lab Scientific Industries, Mumbai, India) was used to measure the temperature stability of the emulsions at varying temperatures. The pH of nano-emulsions was calculated using a pH meter (EquipTronic) at room temperature 37 °C.

vi. Application on fabric:

The prepared nano-emulsions have been applied on organic cotton using two methods:

(i) Continuous process (Padding method): For organic cotton of both 20 and 60 g/m^2 fabrics, sample size 21 cm × 30 cm was used. The fabric has been padded using a 2-dip and 2-nip method at 75% expression (the rate at which fabrics passes through) of the padding mangle. The padded fabric is then dried at 80 °C for five minutes and cured at 110 °C for three minutes. Padding has been carried out for all the five ratios of nano-emulsion for both herbal oils.

(ii) Batch process (Exhaust method): The exhaust has been carried out using a Rota dryer machine. The fabric samples are kept in exhaustion at 1:50 LMR (liquid to material ratio) at 60 °C for one hour. The fabric after the exhaust method has been dried in air at room temperature. Exhaust has been carried out for all the five ratios of nano-emulsion for both the herbal oils.

2.3. Fabric Characterisation

Fabric surface morphology was examined using scanning electron microscopy (SEM) Carl Zeiss Supra 40 VP (Oberkochen, Germany), and samples were analysed under variable pressure conditions at a chamber pressure of 30 Pa. A backscattered electron detector was used at an acceleration voltage of 20 kV to obtain images of the samples. Low magnification images (100× and 50×) were obtained at a working distance of approximately 25 mm, whilst higher magnification images (1000×) were obtained between 5 and 6 mm. In addition, energy dispersive X-ray spectroscopic analysis (EDX) was carried out using Apollo 40SDD (Tilburg, The Netherlands) and is used to determine the elemental composition of fabric treated with nano-emulsions. EDX analysis was performed using an acceleration voltage of 20 kV and a working distance of approximately 15 mm.

The chemical structure of finished and unfinished cotton fabric samples was characterised using Brucker Alpha II attenuated total internal reflectance—Fourier transform infrared spectroscopy (ATR-FTIR, Bruker Daltonik GmbH, Bremen, Germany). The FTIR spectra for cotton woven fabrics were recorded from 4000 to 600 cm^{-1}. The resolution used is 4000 to 550, made up of 32 scans. The internal reflection element of the ATR crystal was diamond.

2.4. Antimicrobial Tests

Antimicrobial activity was evaluated by quantitative (AATCC 100:2019) and qualitative methods (AATCC 147:2016) [51,52]. *Staphylococcus aureus* strain No. ATCC 6538 (Gram-positive bacteria) and *Escherichia coli* strain No. ATCC 10,799 (Gram-negative bacteria) were used for this study. The controls used in the study are unfinished fabrics. The strains were cultured in nutrient agar and sterilised using an autoclave. The control and test samples were allowed in contact with bacteria for 24 h, and the percentage reduction of micro-organisms was determined using the formula:

$$R = (B - A/B) \times 100\% \tag{1}$$

where,

- A denotes the number of bacteria recovered from inoculated treated specimen after 24 h;
- B denotes the number of bacteria recovered from the inoculated treated specimen immediately after inoculation, i.e., 0 h.

In the parallel streak method, the test specimen (rectangular specimens cut 25–50 mm) and control sample is placed in contact with the nutrient agar for 24 h. After incubation, a clear zone of inhibition in mm is calculated using the formula $W = (T - D)/2$, where W is the width of the clear zone of inhibition, T is the total diameter of the test specimen and clear zone in mm, and D is the diameter of the test specimen. A zone of inhibition underneath and around the sample indicates antibacterial resistance. Antifungal tests were conducted using test strain *Aspergillus niger* (strain No. ATCC 6275) to evaluate the antifungal resistance of the finished samples. Samples were incubated for six days in the humidity chamber at 28 °C and 90% relative humidity (AATCC 30: III- 2013) [53].

Tensile strength was evaluated in warp and weft directions using ASTM D 5035-11 (2019) with a gauge length of 75 mm [54]. Wash tests were carried out in accordance with the ISO 2 procedure (IS 15370: 2005) [55]. A launderometer was used with a standard reference detergent (IEC), maintaining a wash temperature of 60 °C, using a material: liquid ratio of 1:50. This was followed by rinsing, washing and drying process. Air permeability was measured to determine the airflow perpendicular through the fabric using a Shirley Air Permeability tester with a pressure drop of 100 Pa (10 mm head of the water column) and with a surface test area of 5.0 cm^2.

3. Results and Discussions

The selection of herbal combinations (*Moringa oleifera, Aegle marmelos*, curry leaf, coconut oil) was based on an extensive survey of herbal constituents, and the proportions of each constituent were based on several trials. As described in Section 2.2, fresh leaves of *Moringa oleifera* and *Aegle marmelos* were used to prepare different herbal extracts and to this extract, curry leaves and coconut oil were added to produce the two combinations—nano-emulsion 1 (*Moringa oleifera*, curry leaf and coconut oil) and nano-emulsion 2 (*Aegle marmelos*, curry leaf and coconut oil). The following section discusses the characterisation of herbal nano-emulsion, including particle size analysis, the thermal stability of herbal nano-emulsions, pH optimisation, determining the percentage add-on of nano-emulsions finish on to the organic cotton fabrics, and determination of Whiteness Index of finished organic cotton fabrics.

3.1. Particle Size Analysis

The particle size analysis study was carried out for "soon after preparation" and after one and two weeks to study the stability of nano-emulsions (Figure 3). The results indicated that the particle size decreases with the increase in the 'oil to surfactant' ratio. Nano-emulsion 2 had a marginal smaller nanoparticle size compared to nano-emulsion 1. There is a shift in particle size observed for Moringa emulsions ranging from 210 nm 'soon after preparation' to 74 nm for the ratio 1:1 after two weeks. This was due to the active phytoconstituents action (the antioxidants present in both the herbs serve to reduce the particle size). The antioxidants are usually present as free radicals in the nano-emulsions and are unstable by nature. Thus enabling the reduction of the particle size of the nano-emulsions.

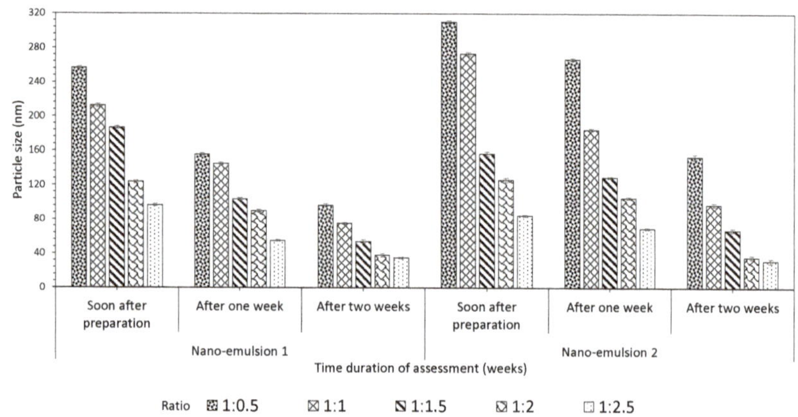

Figure 3. Particle size assessments.

Furthermore, the surfactant being proportionally more prominent than the oil and its interactions with the oil could undergo saponification with fatty acid molecules to form small micelles (aggregate of molecules). Hence, the surfactant, along with antioxidants play a role in the reduction of particle size of the nano-emulsions. Besides, the presence of antioxidants gives better durability of the finish on the fabric since it prevents rancidity.

A similar trend was observed for nano-emulsion 2, where the particle size decreased from 273 nm soon after preparation to 98 nm after two weeks. It is worth mentioning that across all the ratios of both the nano-emulsions, the particle size decreased. For instance for the 1:1 ratio, the particle size variation was 32% between soon after preparation and one week; after two weeks, there was a 65% variation in nanoparticle size. The particle size was prominent during the initial stages in the 100–250 nm region and 80–300 nm for nano-emulsion 1 and 2. In this study, the nano-emulsions were used after one week for treating the organic cotton, where the particle size was in the region 100–150 nm. Interestingly, the particle size of both the nano-emulsions after 60 weeks did not show any variation beyond 30–35 nm (data not presented here), showing excellent stability and shelf-life. Particle size analysis of hibiscus, curry leaf, and coconut oil—herbal extract combination was good in the various ratios, particularly for 90/10 (herbal oil/water), indicating good antimicrobial potential [56]. All the nano-emulsion ratios have been experimented with (1:0.5 to 1:2.5). After optimisation of the nano-emulsion, the 1:1 ratio was found to be appropriate and better than the other ratios, since the oil to surfactant ratio were in equal proportions. The surfactant proportion, when increased, gives the following properties, (1) the ease of penetration into the fabric; (2) particle size decreases; (3) good shelf-life; and (4) the ease of application using both methods (continuous and batch process)—the smaller the particle size, the better the penetration of nano-emulsion into the fabric.

3.2. pH Optimisation of Nano-Emulsions

The optimisation of pH was carried out for the emulsions of all the herbal ratios. 0.1% sodium hydroxide and 0.1% hydrochloric acid solution was added to the emulsions to verify the stability of the nano-emulsion solution. From Figure 4, it can be observed that there is a gradual increase in pH value as the herbal ratio is varied from 1:0.5 to 1:2.5 and is applicable to both the nano-emulsions. For both the nano-emulsions 1 (*Moringa oleifera*) and nano-emulsions 2 (*Aegle marmelos*), the pH of the nano-emulsions was optimised between 5 and 6.

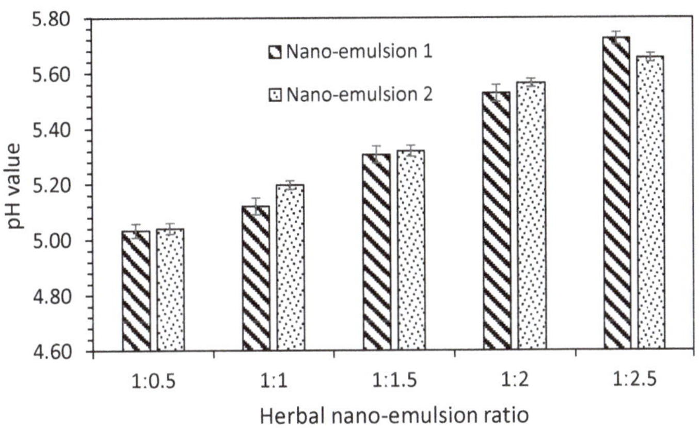

Figure 4. pH value of herbal nano-emulsions.

3.3. Thermal Stability

The thermal stability of the emulsions was carried out by keeping the emulsions in the water bath with a maximum temperature of 95 °C. It was observed that as the temperature of the mixture was increased to 60 °C, the emulsion was no longer stable, indicating the emulsion was stable only up to 60 °C. As the ratio (surfactant to oil) proportion increased, the thermal stability increased to >60 °C. Thus, the thermal stability of nano-emulsions was monitored for three months, and it was stable up to 60 °C throughout this period. However, any further increase in temperature above 60 °C resulted in the breaking of emulsions (oil particles accumulated as a bottom layer). This was observed using a separating funnel. Therefore, nano-emulsion 1 was stable in the region 50–60 °C, whilst nano-emulsion 2 was stable in the region 50–69 °C. The surfactant being a polysorbate reduces the surface tension of the nano-emulsion, increases the shelf life and the thermal stability, as already observed in Figure 5, for both the nano-emulsions. The surfactant–oil interaction is a continuous process of emulsification and saponification, thus reducing the particle size and lowering the surface tension.

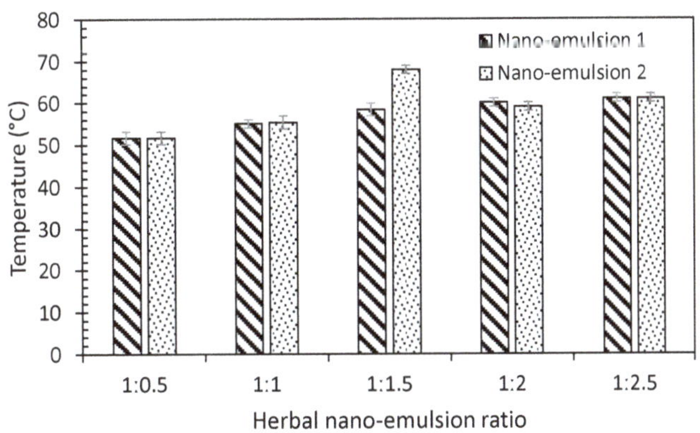

Figure 5. Thermal stability of herbal nano-emulsions.

3.4. Nano-Emulsion Percentage Add-on

The percentage add-on was evaluated for both the nano-emulsions finished fabrics as shown in Figure 6. 20 gsm fabric had a higher add-on than 60 gsm for all nano-emulsions and is explained as follows: (1) The structure of the 20 gsm plain fabric was more open compared with the 60 gsm closely twill weave. The cover factor for 20 gsm in warp direction was 21 whilst for 60 gsm was 63, indicating that area covered by a set of threads in 60 gsm was three times higher than 20 gsm, indicating a tightly woven packed structure of 60 gsm than 20 gsm fabric. (2) This was why 20 gsm had a higher percentage add-on. In addition, it indicates the interaction between the nano-emulsion and the substrate. Had it been anything but mechanical adsorption, 60 gsm would have shown a higher add-on than 20 gsm because the surface area of 60 gsm was higher than 20 gsm, as shown in Table 1. (3) Therefore, the interaction between the nano-emulsion and fabrics was simple mechanical adsorption subject to the voids present in the fabric structure. With a further increase in the area of the fabric, the percentage add-on remained unchanged, indicating that mechanical adsorption took place without any reaction between the nano-emulsions and fabric.

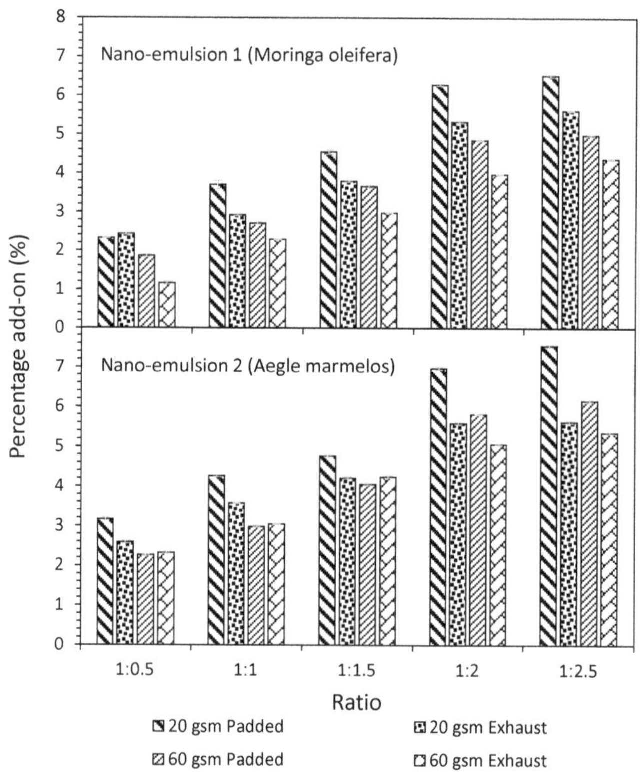

Figure 6. Percentage add-on of nano-emulsions.

Since the mechanism of finishing of fabrics with nano-emulsion is based on mechanical adsorption with no crosslinking, the type of fabric (weave) does not influence resultant properties, as it only depends only on the voids or interspaces present between the fibres/yarns within the fabric. Therefore, fabric weights chosen in this study was to determine the effect of finishing on different lightweight fabrics for its percentage add-on and durability of the finish (in different weave patterns).

Table 1. Physical properties of fabrics.

–	Fabric 1	Fabric 2	Notes
Fabric weight (g/m^2)	20	60	GOTS † certified 100% Organic cotton fabric scoured and bleached white
Thickness (mm)	0.2	0.5	–
Fabric structure	Plain weave (1/1)	Twill weave (4/1) warp faced	1.5 m width
Fabric count EPI × CPI	75 × 27	117 × 38	Ends per inch and picks per inch
Warp yarn count (tex)	8	29	–
Weft yarn count (tex)	10	31	–
Cover factor K = k1(warp) + k2(weft)	21.2 + 64	63 + 21	Area covered by a set of threads
Air permeability (cc/s)	>200	63 (±2.73)	The rate of airflow perpendicular through the fabric
Tensile strength (N) Warp Weft	322.39 (±11.01) 226.01 (± 1.29)	662.09 (±2.71) 287.84 (±12.76)	Force required to break the fabric under tension
Breaking extension (%) Warp Weft	7.8 (±0) 27.53 (± 0.26)	15.68 (±0.14) 14.67 (±0.23)	Extension at peak force

The number in brackets indicates the standard deviation. † GOTS—Global organic testing standards.

For the 1:2 ratio, a higher add-on percentage was observed for both the fabrics (20 and 60 gsm). This could be due to an increase in the surfactant to stabilise the herbal nano-emulsion and disperses into the fibre structure more uniformly. A similar pattern was also observed for nano-emulsion 2 in the continuous (padding method) as well as for the batch process (exhaust method). The 1:2 ratio had a better percentage add-on compared to other ratios. Overall, it can be inferred that the continuous (padded) process had a negligible higher difference than the batch process (exhaust). It is interesting to know that nano-emulsions ratio 1:0.5 to 1:2.5 showed a distinct change in percentage add-on on organic cotton fabrics. When comparing the nano-emulsions of both the herbs, the performance of *Aegle marmelos* was better than *Moringa oleifera* with respect to the preparation of nano-emulsions, the ease of finishing, and absorption higher percentage add-on. In the lower ratio (1:0.5), the surfactant is in a lower quantity, and will not have enough affinity to penetrate the fabric in the continuous method (padding), since the time of contact between the fabric and solution is lesser, whereas the time of contact is higher in the exhaust method. Hence, at a lower concentration of surfactant, there is a slight increase in the percentage add-on in the exhaust method compared with the padding method. As the surfactant concentration increases, the strike of the finish on cotton increases due to the lower surface tension of the emulsion. Hence, the padding method can adsorb more and increase the percentage add-on compared to the exhaust method.

The three replicates from the experiments showed a slight variation for different ratios for both the fabrics and nano-emulsions. Therefore, the error bars were low, as seen in Figure 6. In addition, the fabrics were finished in a single batch of 100 m length showing the uniform application of the percentage add-on, and the nano-emulsions prepared were consistent throughout this study. This could be the reason for the low variation in the samples.

3.5. Whiteness Index

The change in the Whiteness Index for nano-emulsion 1, 20 and 60 gsm fabrics increased as the herbal ratio varied from 1:0.5 to 1:2.5. Interestingly, the change in Whiteness Index for 20 gsm fabric for nano-emulsion 1, particularly a 1:1 ratio, was approximately

12%, whilst for 60 gsm fabric, this change was approximately 9.4%. The change in th Whiteness Index was distinct for nano-emulsion 1 across all the ratios. In nano-emulsion the change in the Whiteness Index increased marginally when the ratio varied from 1:0. to 1:2.5. However, at the 1:1 ratio, the change in the Whiteness Index for 20 and 60 gsm was 7.8% and 6.1%, respectively. This also reveals that the change in the Whiteness Inde was well below 10% for nano-emulsion 2 compared with nano-emulsion 1. The Whitenes Index for 20 gsm fabric for both the nano-emulsions were marginally higher than 60 gsm fabric. This was mainly due to the finer yarn linear density and open structure of the plai weave (20 gsm) than 60 gsm fabric. The penetration of nano-emulsions was higher fo 20 gsm as the fabric structure was open, resulting in a distinct change in the Whitenes Index. The Whiteness Index above 10% is visually noticeable (to the naked eye) compare to Whiteness Index below 10%, which is difficult to notice. This is shown in a blue dotte line, Figure 7.

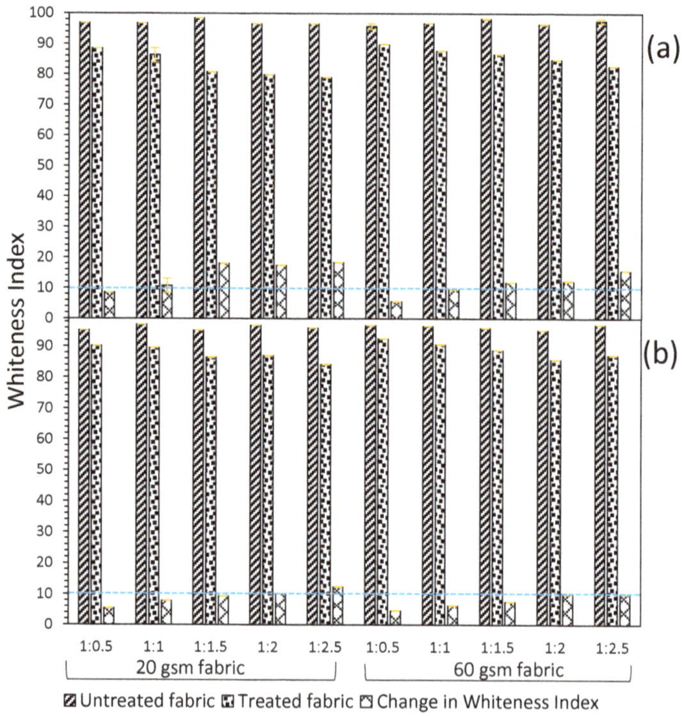

Figure 7. Whiteness Index for (**a**) Nano-emulsion 1 (*Moringa oleifera*); (**b**) Nano-emulsion 2 (*Ae gle marmelos*) both 20 and 60 gsm organic cotton (dotted blue line indicates 10% Whiteness Index tha is noticeable)—padding method.

3.6. Physical Properties

The physical properties of both the 20 and 60 gsm fabrics are presented in Table 1. could be observed 20 gsm fabric has a plain weave structure, whilst 60 gsm fabric is a twi structure that has a marginally higher fabric count. Both the fabrics have a higher tensil strength in the warp direction compared to the weft direction, and this can be attributed to a higher number of warp yarns and finer yarn linear density. Organic cotton was scoure and bleached. These two fabrics were finished with two different nano-emulsions, 1 and 2 using continuous (padding) and batch methods (exhaust).

3.7. Antimicrobial Assessments

3.7.1. Quantitative Tests

The quantitative antimicrobial assessments reveal that both the herbal oil nano-emulsions have excellent resistance to Gram-positive and Gram-negative bacteria. The percentage reduction of micro-organisms (R%) is shown in Table 2. For 20 gsm for herbal nano-emulsion 1, the percentage reduction was in the range 98.44–99.30%, and for nano-emulsion 2, the range was between 99.08% and 99.72%. However, in 60 gsm fabric, the range for nano-emulsion 1 and 2 was 99.02–99.29% and 99.74–99.87%, respectively. Thus, the reduction of micro-organisms for 60 gsm was marginally higher for nano-emulsion 2 than nano-emulsion 1. This could also be attributed to the fabric structure of 60 gsm fabric (twill weave) and fabric thickness, which had marginally higher pick-up of nano-emulsions, which offered higher resistance than 20 gsm plain-woven fabric.

Table 2. Reduction of micro-organisms (AATCC 100)—continuous process (padding method).

Herbal Ratio	Test Culture	Reduction of Microorganisms (R) (%) Continuous Process (Padded)			
		Moringa oleifera		*Aegle marmelos*	
		20 gsm	**60 gsm**	**20 gsm**	**60 gsm**
1:0.5	*S. aureus*	98.58	99.02	99.72	99.84
	E. coli	98.46	99.03	99.70	99.82
1:1	*S. aureus*	98.44	98.55	99.81	99.81
	E. coli	98.13	98.38	99.78	99.80
1:1 After 10 washes	*S. aureus*	98.17	98.28	98.21	99.72
	E. coli	97.79	98.14	98.09	99.66
1:1 After 20 washes	*S. aureus*	97.34	97.41	97.70	99.26
	E. coli	96.86	97.16	97.58	99.03
1:1.5	*S. aureus*	98.91	99.21	99.08	99.76
	E. coli	98.84	99.15	99.16	99.72
1:2	*S. aureus*	98.82	98.80	99.25	99.77
	E. coli	98.53	98.86	99.24	99.74
1:2.5	*S. aureus*	99.30	99.29	99.69	99.87
	E. coli	99.05	99.07	99.70	99.76

The standard deviation values for both the fabrics, nano-emulsions and for all ratios were negligible in the range 0.00006–0.00026; variations between the repeat samples were minimum.

The washing tests on the finished fabrics were carried out on the 1:1 ratio as a representative sample. 1:1 ratio (equal proportion of oil and surfactant) showed optimum values across a range of physical properties, and the results obtained from this ratio was representative of the remaining samples. Hence, the 1:1 ratio was chosen for evaluating the wash fastness after 10 and 20 wash cycles and antibacterial tests were reported. In the wash durability (padded method), after 10 and 20 washes for 20 gsm 1:1 ratio, there was a marginal decrease in the reduction of antimicrobial resistance compared to before wash, against *S. aureus* 0.27% and 1.1% respectively for nano-emulsion 1. Similarly, for nano-emulsion 1, 20 gsm fabric, there was a similar trend of marginal decrease in the reduction of antimicrobial resistance against *E. coli* ranging from 0.35% for ten washes to 1.29% for 20 washes. In the 20 gsm 1:1 ratio, after ten washes for nano-emulsion 2, the reduction of micro-organisms decreased to 1.57% against *S. aureus*, 1.69% *E. coli* compared to before wash. There was a further reduction in the antimicrobial resistance after 20 washes, 1.69% against *S. aureus*, 2.2% against *E. coli*.

For 60 gsm fabric with nano-emulsion 1, a decrease in the antimicrobial resistance between before and after wash was observed, and these were marginal with 0.1% against *S. aureus*, 0.24% *E. coli* (10 washes) and 0.98% against *S. aureus* and 1.24% against *E. coli*

(20 washes). For nano-emulsion 2 for both 10 and 20 washes, the decrease in the reduction of micro-organisms were less than 1%. Hence, it can be inferred that 60 gsm fabric outperformed 20 gsm fabric with regard to antimicrobial resistance. In addition, the decrease in antimicrobial resistance after 20 wash cycles was minimum for both the fabrics and nano-emulsions, indicating that good finishing was achieved. The finishing of fabrics with nano-emulsions uniformly has a substantial effect on the overall antimicrobial resistance of fabrics. Recently, cotton fabrics were finished using *Ocimum sanctum* (*Holy Basil or Tulsi*) using the padding method with an average particle size of 33.2 nm and showed antimicrobial resistance to *E. coli*, *S. aureus* and antifungal resistance [40]. The authors reported that herbal nano-emulsions inhibited the growth of micro-organisms due to smaller size and uniform coating on the finished cotton fabrics, and good wash durability after 30 wash cycles [57].

The fabrics were also finished using the exhaust method (Table 3), and their antimicrobial resistance was evaluated for 1:1 and 1:2 ratios. It can be inferred that 60 gsm fabric is marginally better than 20 gsm fabric against both the Gram-positive and Gram-negative bacteria. Wash tests also revealed there was a marginal drop in the reduction of micro-organisms after 10 and 20 washes; however, the drop in the antimicrobial resistance was lower than the padded method, implying the exhaust method treated fabric has marginally better antimicrobial resistance compared to the padded treated fabric. Appendix A illustrates antimicrobial assessments (AATCC 100) for 20 and 60 gsm fabrics taken at 0 and 24 h.

Table 3. Reduction of micro-organisms (AATCC 100)—Batch process (exhaust method).

Herbal Ratio	Test Culture	Reduction of Microorganism (R) %—Exhaust Method			
		Moringa oleifera		*Aegle marmelos*	
		20 gsm	60 gsm	20 gsm	60 gsm
1:1	*S. aureus*	98.76	99.21	99.04	99.74
	E. coli	98.60	99.28	99.13	99.69
1:1 After 10 washes	*S. aureus*	98.39	99.09	98.67	99.18
	E. coli	97.82	99.04	98.50	99.17
1:1 After 20 washes	*S. aureus*	97.62	98.26	98.28	98.52
	E. coli	97.57	98.15	97.70	98.26
1:2	*S. aureus*	99.43	99.12	99.27	99.78
	E. coli	99.08	99.03	99.04	99.78

Gas chromatography mass spectroscopy (GC-MS) and GC (gas chromatography) analysis will enable to identify the chemical compounds and phytoconstituents present in *Moringa oleifera*, *Aegle marmelos*, *Murraya koengii* (curry leaf), and coconut oil and is highlighted in Appendix A.1. All the above three herbal oils (*Moringa oleifera*, *Aegle marmelos*, *curry leaf*) contains oleic acid components (dodecanoic, tetradecanoic acid, hexadecenoic, octadecanoic acid and oleic acids, [58]. Coconut oil has 92% saturated fatty acids [48,49]. GC-MS analysis of coconut oil revealed various fatty acids, among them capric acid, lauric acid, myristic acid and palmitic acid, were in high proportions. Thus, C_{16}–C_{20} fatty acids found in these herbs and the phytoconstituents present demonstrated potential antibacterial and antifungal properties, as reported in previous research, Appendix A.1. These phytoconstituents, chemical compounds, and fatty acids inhibit the growth of bacteria, microbes, and fungi. Since the constituents of these three herbs are similar to each other (Appendix A.1), it offers a synergistic effect when blended and retaining the properties of each constituent. The antimicrobial values obtained on finishing cotton fabrics with the nano-emulsions 1 and 2 in this study justifies the above statement.

3.7.2. Qualitative Tests

The antimicrobial assessments using the parallel streak method (AATCC 147:2016) are illustrated in Figure 8. For 20 gsm fabric, there was no growth of bacteria below the fabric for both the nano-emulsions 1 and 2. However, for the 60 gsm fabric, there was a clear zone of inhibition, and no growth below the fabric was observed. Furthermore, this inhibition zone was marginally higher against *S. aureus* nano-emulsion 2 than *E. coli*. This pattern was also noticed for nano-emulsion 1, indicating that 60 gsm fabric has better antimicrobial resistance for Gram-positive bacteria (*S. aureus*) than Gram-negative bacteria (*E. coli*).

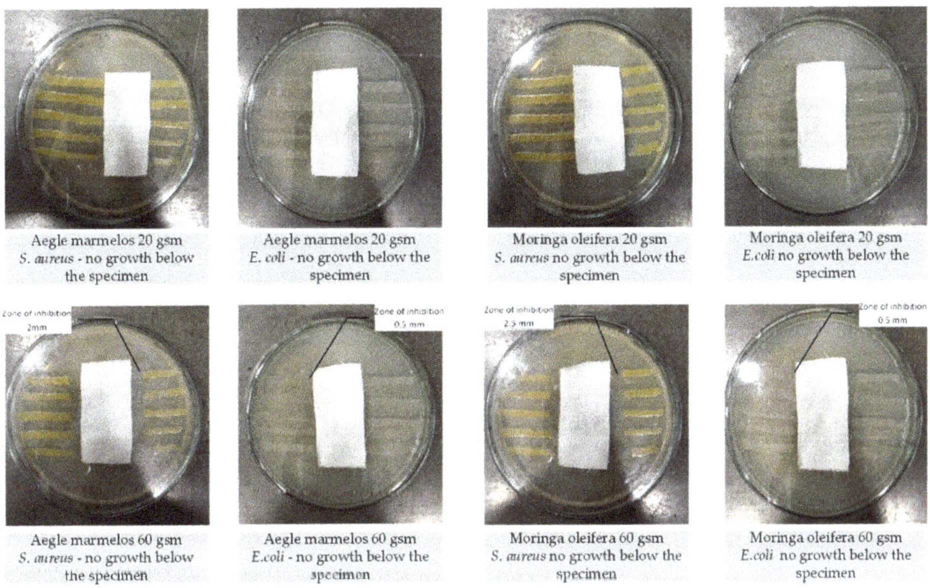

Figure 8. Antimicrobial tests AATCC 147:2016 parallel streak method.

3.7.3. Antifungal Tests

Fabrics treated with nano-emulsions 1 and 2 using the continuous (padding) method was presented here as a representative sample. For 60 gsm fabric, there was no growth of fungi (*Aspergillus niger*), and a clear zone of inhibition was observed, −42 mm for nano-emulsion 1 and 45 mm for nano-emulsion 2. However, there was a 10% growth of fungi for both the nano-emulsions 1 and 2 when using 20 gsm fabric (Table 4). These tests indicated that 60 gsm fabric showed excellent antifungal resistance for both the nano-emulsions.

Table 4. Antifungal test AATCC 30-III 2013.

Sample Identification		Zone of Inhibition (Continuous Method)		Rating *	Interpretation
Fabric	Ratio	*Moringa oleifera*	*Aegle marmelos*		
20 gsm	1:1	No zone of inhibition	No zone of inhibition	1	No zone of inhibition could be seen around the fabric. There was a trace of fungal growth on the samples
60 gsm	1:1	42 mm	45 mm	0	A zone of inhibition can be seen. Antifungal activity present

* Rating description: 0: no growth; 1: trace of growth (<10%); 2: light growth (10–20%); 3: medium growth (30–60%) 4: and heavy growth (60% to complete coverage).

3.8. ATR-FTIR Characterisation and SEM Analysis

It can be observed from the FTIR, Figures 9 and 10, the functional groups of the cotton cellulose remain unaffected between treated and untreated fabrics. The peaks observed for 20 and 60 gsm are given below. FTIR analysis reveals that there was no chemical change in the structure of organic cotton, indicating that the herbal finishing of fabrics did not affect the cellulose polymer structure.

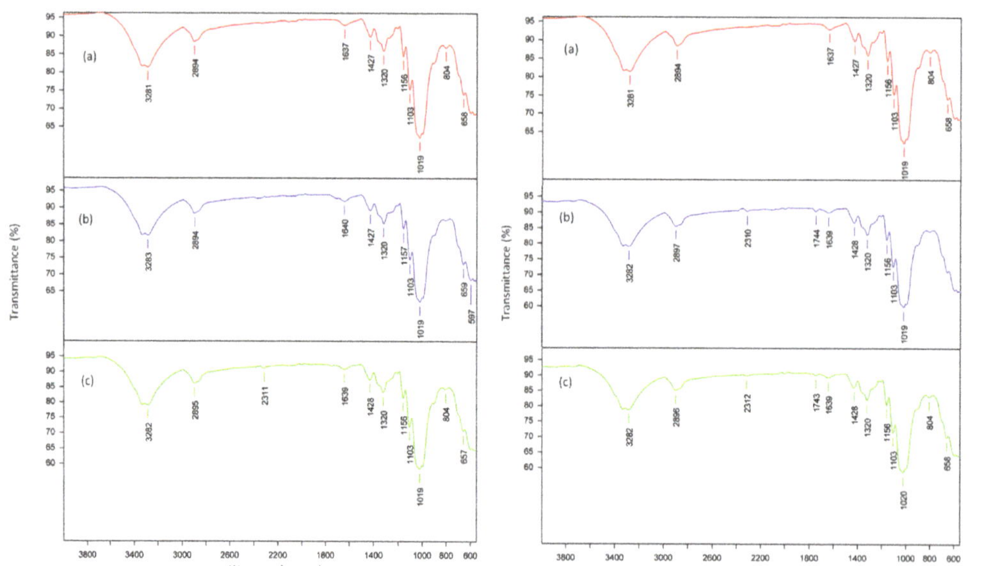

Moringa oleifera 1:1 20 gsm (a) control; (b) exhaust and (c) padding   Aegle marmelos 1:1 20 gsm (a) control; (b) exhaust and (c) padding

Figure 9. FTIR for *Moringa oleifera* and *Aegle marmelos* 20 gsm fabrics.

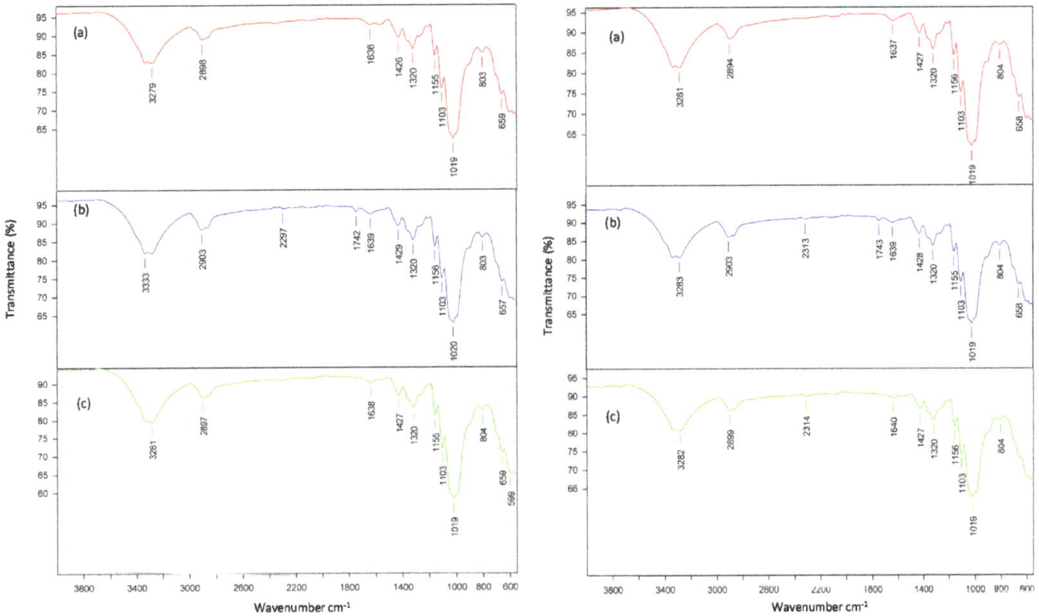

Moringa oleifera 1:1 60 gsm (a) control; (b) exhaust and (c) padding Aegle marmelos 1:1 60 gsm (a) control; (b) exhaust and (c) padding

Figure 10. FTIR for *Moringa oleifera* and *Aegle marmelos* 60 gsm fabrics.

In the case of control organic cotton fabric for 20 gsm, the spectra showing assignments at 3281 cm^{-1} relating to O–H stretching (Carboxylic acid), 2894 cm^{-1} O–H stretching (alcohol), 1637 cm^{-1} C–H bending, 1427 cm^{-1} O–H bending, 1320 cm^{-1} O–H bending (phenol group), 1156 cm^{-1} C–O–C stretching. In the case of 60 gsm unfinished fabric, organic compounds present were identified as follows: 3279 cm^{-1} O–H stretch (alcohol), 2898 cm^{-1} N–H amine salt, 1636 cm^{-1} C=C stretching, 1426 cm^{-1} O–H bending, 1320 cm^{-1} O–H bending, 1155 cm^{-1} C–O stretching.

The organic compounds present in 20 gsm fabric treated with (*Moringa oleifera*) nano-emulsion 1, by exhaust method as observed from Figure 9, assignments include 3283 cm^{-1} C–H stretching, 2894 cm^{-1} C–H stretching (alkane), 1640 cm^{-1} C–N stretching, 1427 cm^{-1} O–H bending, 1157 cm^{-1} C–O–C stretching, and 1103 cm^{-1} C–N stretching. While the fabrics finished by the continuous method for 20 gsm consists of the following assignments 3282 cm^{-1} O–H stretching, 2895 cm^{-1} C–H stretching (alkane), 1639 cm^{-1} C=C stretching (alkene), and 1428 cm^{-1} O–H bending. The treated fabrics with nano-emulsion 1 showed spectra assignments similar to untreated fabrics.

Similarly, according to Figure 10, the organic compounds present in *Moringa oleifera* for 60 gsm treated samples by exhaust method include 3333 cm^{-1} N–H stretching, 2903 cm^{-1} C–H stretching (alkane group), 1742 cm^{-1} C=O stretching (ester group), 1639 cm^{-1} C=C stretching (alkene), 1429 cm^{-1} O–H bending, 1320 cm^{-1} O–H bending (phenol), 1156 cm^{-1} C–O–C stretching, and 1103 cm^{-1}, C–N stretching. For 60 gsm fabric using continuous process treatment, the spectra assignments include 3281 cm^{-1} O–H stretching, 2897 cm^{-1} OH stretch (alcohol), 1638 cm^{-1} C=C stretching, 1427 cm^{-1} O–H bending, 1320 cm^{-1} O–H bending, and 1103 cm^{-1} C–N stretching. It is observed from the FTIR analysis that there is an insignificant change in the peaks of ATR spectra of the finished fabric with the herbal oils as compared with the unfinished fabrics. This phenomenon was due to the simple mechanical adsorption of the oil in the voids of the fabric and held by Van der Waal forces. There is no chemical bond (ionic and covalent) formation between the fabric and oil. The nano-emulsions possess a high affinity toward the cotton fabric and penetrate easily into the voids of the fabric structure, and the steric hindrance of the oil molecule prevents it from leaving the fabric surface easily, resulting in good durability of the finish on the cotton fabric.

ATR spectra assignments for oil mixture 1 (*Moringa oleifera*) (Figure 11) include 2922 cm^{-1} O–H stretching; 1742 cm^{-1} C–H bending, and 1157 cm^{-1} C–O–C stretching; however, for the cotton fabric ATR assignments include 3281 cm^{-1} relating to O–H stretching (Carboxylic acid), 2894 cm^{-1} O–H stretching (alcohol), 1637 cm^{-1} C–H bending, 1427 cm^{-1} O–H bending, 1320 cm^{-1} O–H bending (phenol group) and 1157 cm^{-1} C–O–C stretching. These ATR spectra for cotton fabrics are in line with the previously reported work [59]. The ATR spectra for the finished fabrics, as shown in Figure 9c, compared with the unfinished fabric Figure 9a, show insignificant changes in the peaks. However, it is observed that oil mixture 1 has a C–O–C stretching and C–H bending, which were similar to the unfinished and finished cotton fabrics (Figure 11). There is an absence of characteristic peaks of the oil functional group due to the nanoparticle size of the oil mixture. Thus, confirming that there is no chemical interaction following the finishing of organic cotton fabrics with the herbal oil mixture. The above phenomenon was similarly observed for oil mixture 2 (*Aegle marmelos*) and for the finished cotton fabrics with oil mixture 2.

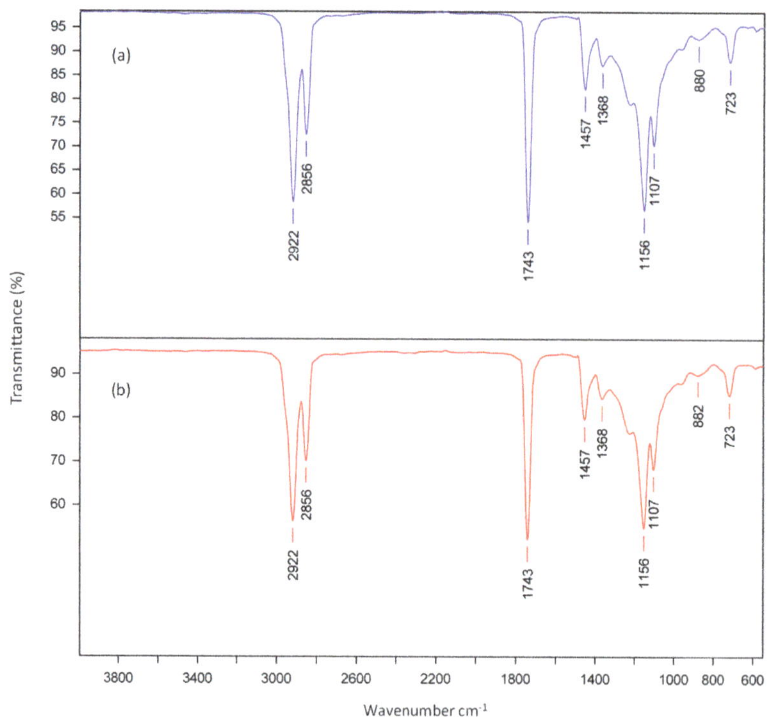

Figure 11. FTIR for (**a**) *Moringa oleifera* oil mixture and (**b**) *Aegle marmelos* oil mixture.

ATR spectra for herbal oil mixture 1 (*Moringa oleifera*, curry leaf and coconut oil) and mixture 2 (*Aegle marmelos*, curry leaf and coconut oil) the assignments were similar (2922 cm^{-1} O–H stretching (alcohol); 1742 cm^{-1} C–H bending; and 1157 cm^{-1} C–O–H stretching).

SEM analysis (Figure 12) for 20 and 60 gsm fabrics reveal that as the herbal ratio varied from 1:1 to 1:2.5 ratio, the ribbon-like shaped cotton fibres gradually flattened, and striations can be noticed on the fibre surface due to treatment with both the nano-emulsions. These visible changes in fibre morphology can be noticed for 60 gsm fabrics and were marginally higher when treated with nano-emulsion 2. EDX analysis showed the presence of carbon elements for both the fabrics, indicating no other chemical constituents after treatment with both the nano-emulsions. However, there were minor traces of silicon and aluminium present, as shown in Figure 12. These elements arise during the processing of finishing fabrics with the nano-emulsions, where the nip rollers were coated with silicone rubber, and wet-pickup rollers were made of aluminium. This resulted in the samples showing traces of silicon and aluminium compounds, as seen in EDX analysis (Figure 12).

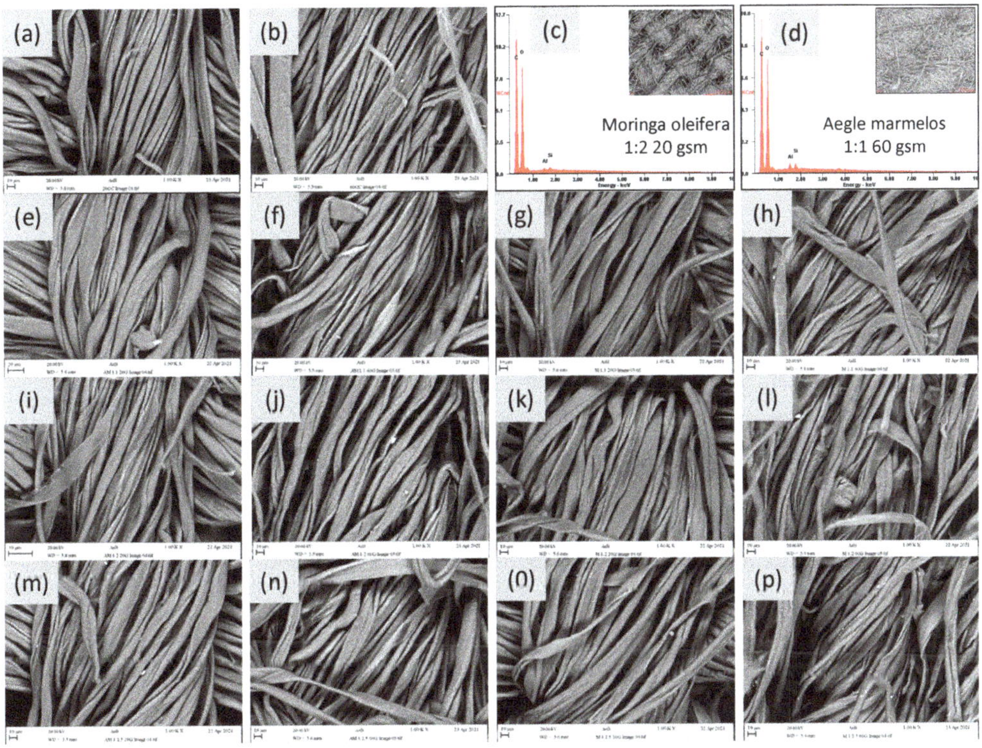

Figure 12. SEM of treated fabrics (padding method); (**a**) 20 gsm control fabric; (**b**) 60 gsm control fabric; (**c**) EDX analysis 1:2 nano-emulsion (*Moringa oleifera*) 20 gsm fabric; (**d**) EDX analysis 1:1 nano-emulsion 1 (*Aegle marmelos*) 60 gsm fabric; (**e**) 20 gsm treated with nano-emulsion 2, 1:1 ratio; (**f**) 60 gsm treated with nano-emulsion 2, 1:1 ratio; (**g**) 20 gsm treated with nano-emulsion 1, 1:1 ratio; (**h**) 60 gsm treated with nano-emulsion 1, 1:1 ratio; (**i**) 20 gsm treated with nano-emulsion 2, 1:2 ratio; (**j**) 60 gsm treated with nano-emulsion 2, 1:2 ratio; (**k**) 20 gsm treated with nano-emulsion 2, 1:2 ratio; (**l**) 60 gsm treated with nano-emulsion 1, 1:2 ratio; (**m**) 20 gsm treated with nano-emulsion 2, 1:2.5 ratio; (**n**) 60 gsm treated with nano-emulsion 2, 1:2.5 ratio; (**o**) 20 gsm treated with nano-emulsion 1, 1:2.5 ratio and (**p**) 60 gsm treated with nano-emulsion 1, 1:2.5 ratio.

3.9. Tensile Strength

Tensile strength of both the fabrics (20 and 60 gsm) decreased when the herbal ratio increased from 1:0.5 to 1:2 in warp and weft directions for both continuous and batch processes (Table 5). For 60 gsm fabric, the tensile strength in the warp direction gradually decreased as the herbal ratio varied from 1:0.5 to 1:2 for both the nano-emulsions. However, with nano-emulsion 2, this decreases in tensile strength—warp and weft directions (for the continuous and batch process) were lower than finishing with nano-emulsion 1. The cotton fabrics are made of 93% cellulose. Generally, there are two mechanisms by which cotton fabrics can be finished (1) forming a monolayer on the fabric surface; (2) crosslinking method. The herbal nano-emulsion used in this study has a hydrophobic nature and forms a monolayer of oil on the surface of the fabric without the crosslinking formation. The surface finishing of cotton fabric is due to the mechanical adsorption of nano-emulsions and depends on the pressure applied during the padding mangle process. The finishing c cotton fabrics with nano-emulsion neutralises the zeta potential on the surface of the fabri (being cationic in nature). This increases the penetration of the solution onto the fabri resulting in a minor reduction of tensile strength of fabrics. It is also worth highlightin that the pH of the nano-emulsions was 5.0–6.0 (acidic) and cotton is sensitive to acid therefore affecting the tensile strength of both fabrics to an extent. However, it is wort mentioning that generally, the tensile strength of the fabric is affected during any finishin of textiles, and in this study, the overall strength of fabrics possesses sufficient strengt The breaking extension of the finished fabrics is shown in Appendix B.

Table 5. Tensile strength (N) *Moringa oleifera*.

I	20 gsm Padding		60 gsm Padding		20 gsm Exhaust		60 gsm Exhaust	
	Warp	Weft	Warp	Weft	Warp	Weft	Warp	Weft
Control	322.39 ± 11.01	226.01 ± 1.29	662.09 ± 2.71	287.84 ± 12.76	322.39 ± 11.02	226.01 ± 1.29	662.09 ± 2.71	287.84 ± 2.76
1:0.5 ratio	317.61 ± 21.18	220.59 ± 2.77	609.42 ± 74.43	265.35 ± 15.85	274.92 ± 32.76	198.37 ± 7.25	604.91 ± 66	254.26 ± 2.68
1:1 ratio	273.99 ± 0.25	217.25 ± 4.38	641.55 ± 20.65	266.18 ± 9.83	306.72 ± 15.90	224.2 ± 1.17	596.43 ± 37.29	231.01 ± 5.33
1:2 ratio	202.64 ± 3.82	195.14 ± 7.41	572.09 ± 31.22	225.61 ± 5.23	239.73 ± 30.91	189.33 ± 32.22	548.09 ± 17.95	267.32 ± 2.96
II	Tensile Strength (N) *Aegle marmelos*							
	20 gsm Padding		60 gsm Padding		20 gsm Exhaust		60 gsm Exhaust	
Control	322.39 ± 11.02	226.01 ± 1.29	662.97 ± 2.71	287.85 ± 12.76	322.39 ± 11.02	226.92 ± 1.29	662.97 ± 2.71	287.85 ± 12.76
1:0.5 ratio	306.43 ± 0	177.56 ± 0	629.93 ± 0	255.85 ± 0	249.42 ± 24.37	217.59 ± 23.33	609.62 ± 28.99	243.6 ± 19.03
1:1 ratio	297.25 ± 10.15	211.51 ± 11.86	599.9 ± 21.34	248.65 ± 9.53	288.78 ± 8.75	206.66 ± 7.0	553.29 ± 11.47	243.8 ± 8.06
1:2 ratio	313.23 ± 0	207.16 ± 0	626.39 ± 0	275.12 ± 0	280.57 ± 18.29	212.63 ± 8.89	529.54 ± 1.33	215.85 ± 6.58

3.10. *Air Permeability*

Air permeability was evaluated for both the fabrics 20 and 60 gsm, and it denote the airflow perpendicular through the fabric, and it depends on the cover factor of fabri (area covered by a set of threads), fabric structure and fabric count (number of warp an weft yarns per unit area). The 60 gsm fabric had a higher cover factor with a higher fabri count, providing resistance to airflow. Hence it can be observed that the rate of airflov was 60 cc/s. The untreated 60 gsm fabric had marginally higher air permeability (63 cc/s than the fabrics treated with nano-emulsions 1 and 2. However, in 20 gsm fabric, th fabric had a plain-woven structure and lower fabric count and cover factor than 60 gsn fabric. As a result, the air permeability for 20 gsm fabric was >200 cc/s for both finishe (nano-emulsions 1 and 2) and unfinished fabric.

4. Conclusions

The market potential for the use of antimicrobial textiles to improve hygiene woul potentially grow and would reach USD 12.3 bn by 2024 at a compound annual growt rate (CAGR) 5.4% between 2019 and 2024 [60]. In addition, the user awareness toward th preference of environmentally friendly antimicrobial hygiene textiles [61] is also expanding indicating the focus toward improving hygiene in an environmentally friendly way. I this study, two organic cotton fabrics (20 and 60 gsm) were finished with two differen herbal nano-emulsions—*Moringa oleifera*, coconut oil with curry leaf (nano-emulsion 1) an *Aegle marmelos*, coconut oil with curry leaf (nano-emulsion 2) at varying ratios using tw different methods of finishing—continuous (padding) and batch process (exhaust). Result indicated that the organic cotton fabrics finished with two nano-emulsions had excellen antimicrobial efficacy against Gram-positive (*S. aureus*) and Gram-negative bacteria (*E. col* with good wash durability. For 60 gsm fabric, when finished with nano-emulsion 2, showe marginally higher antimicrobial resistance in the region 99.74–99.87% for varying herba ratios than 20 gsm fabrics. Wash tests of finished fabrics also showed good antimicrobia resistance; however, the reduction in micro-organisms dropped between 0.27–2.2% fo 10 and 20 washes. The parallel streak method (AATCC 147) also revealed no growth o bacteria below the fabric, and a clear zone of inhibition for 60 gsm fabrics was observe In addition, the finished fabrics (60 gsm) showed a clear zone of inhibition in the rang of 42–45 mm and with no growth of fungi (*A. niger*), demonstrating excellent antifungal efficacy. The particle size evaluations showed that herbal constituents were in the region of 55–150 nm and 70–266 nm for nano-emulsions 1 and 2, respectively. The nano-emulsions were stable and showed further reduction in particle size after two weeks. The percentage add-on for both the nano-emulsions increased with the increasing herbal ratio (1.0.5 to 1:2.5), and the overall add-on for nano-emulsion 1 and 2 was 4–6% and 5–6%, respectively, demonstrating good finishing of cotton fabrics. Optimisation of pH (5.0–6.0) and thermal stability of nano-emulsions (50–60 °C) showed good stability of both the nano-emulsions. Whiteness Index assessment showed that 20 gsm fabric had an Index marginally above 10%, whilst for 60 gsm fabric, the index was well below 10% for both the nano-emulsions. An index below 10% cannot be noticed in the naked eye. This shows that the finished fabrics with nano-emulsions were not affected from their original shade (bleached white), and the tonal changes were marginal.

FTIR analysis for finished and unfinished fabrics (20 and 60 gsm) revealed no chemical interaction on finishing of fabrics with herbal oil mixture, confirming that the finishing of fabrics was due to simple mechanical adsorption of herbal oil in the voids of the fabric structure. Furthermore, an insignificant change in the peaks of the ATR spectra between finished (20 and 60 gsm) and unfinished fabrics confirmed this phenomenon. When examining the tensile strength of both the fabrics and for nano-emulsions 1 and 2, the tensile strength in the warp direction was higher than the weft direction, and as the herbal ratio varied from 1:0.5 to 1:2.5, the tensile strength decreased when compared to unfinished fabric. The finished and unfinished fabrics were evaluated for air permeability, which showed no change in the rate of airflow (>200 cc/s) for 20 gsm; however, for 60 gsm fabric, there was a marginal change in the rate of airflow (60–65 cc/s) across all the herbal ratios. These findings also showed that finished fabrics did not alter the fabric structure, allowing airflow through the fabric. SEM analysis also confirmed marginal variations in the fibre structure between finished and unfinished fabrics (20 and 60 gsm). It is also interesting to observe that the 1:1 herbal ratio for 20 and 60 gsm fabrics among various herbal ratios showed optimum results for percentage add-on, antimicrobial, antifungal efficacy, and good wash durability. Based on the above assessments, the 1:1 herbal ratio using continuous process (padding method) showed excellent results for both fabrics. This was also due to the mechanical adsorption of nano-emulsions onto the fabric surface.

Both the herbal combinations studied in this research (*Moringa oleifera* and *Aegle marmelos*) have demonstrated excellent antimicrobial resistance. Generally, *Moringa oleifera* is widely preferred for skin ailments and other diseases, which was reported by previous studies [62,63] and methanol extracts of *Moringa oleifera* show resistance to a wide range of bacterial strains [64]. This also indicates the potential of these herbs to develop further innovative combinations to treat various diseases and ailments. This research adds to the literature by demonstrating a novel method of preparing nano-emulsions using a combination of herbs and finishing them on organic cotton fabrics. This method of finishing fabrics with all-natural herbal nano-emulsion to develop antimicrobial finishes on cotton fabrics can be widely used for healthcare and hygiene textiles. The herbal finishes also possess low adverse effects compared to synthetic compounds and are environmentally friendly. The study can be further developed with varying concentrations of herbal nano-emulsions on different fabrics and understand the mechanism of processing for the desired end-use.

Author Contributions: Conceptualisation—U.S. and P.D.V.; methodology—U.S.; P.D.V., S.P., and S.K.; data collection and preparation—S.K. and S.P.; investigation—U.S., P.D.V., S.P., and S.K.; formal analysis—P.D.V., S.K., and S.P; writing—original draft preparation—P.D.V. and U.S.; writing—review and editing—P.D.V., U.S., S.P., and S.K.; supervision—U.S.; project administration—U.S. and P.D.V.; funding acquisition—P.D.V. and U.S. All authors have read and agreed to the published version of the manuscript.

Funding: This research received internal funding—Global Challenges Research Fund (GCRF) through Manchester School of Art and Research Centre, Faculty of Arts and Humanities, Manchester Metropolitan University, UK Project ID: 328682.

Institutional Review Board Statement: Not applicable.

Informed Consent Statement: Not applicable.

Data Availability Statement: The data presented in this study are available on request from the corresponding author.

Acknowledgments: The authors would like to acknowledge the kind contribution from the following technical/support colleagues: Hayley Andrews, MMU, UK., Scanning Electron Microscopy, EDX analysis; Derek Hebdon—Textile testing, MMU, UK.; all staff of the Department of Fibres and Textile Processing Technology, ICT, Mumbai.; Mrinal Choudhary, WRA (Wool Research organisation), Smita Bhatt and Seema Patel and Madhura Nerurkar, Calantha Biotech, Mumbai India; and V. Dhanapal, India, for providing support toward development 3D images.

Conflicts of Interest: The authors declare no conflict of interest.

Appendix A.

Antimicrobial Tests (AATCC 100) for 20 and 60 gsm Fabrics.

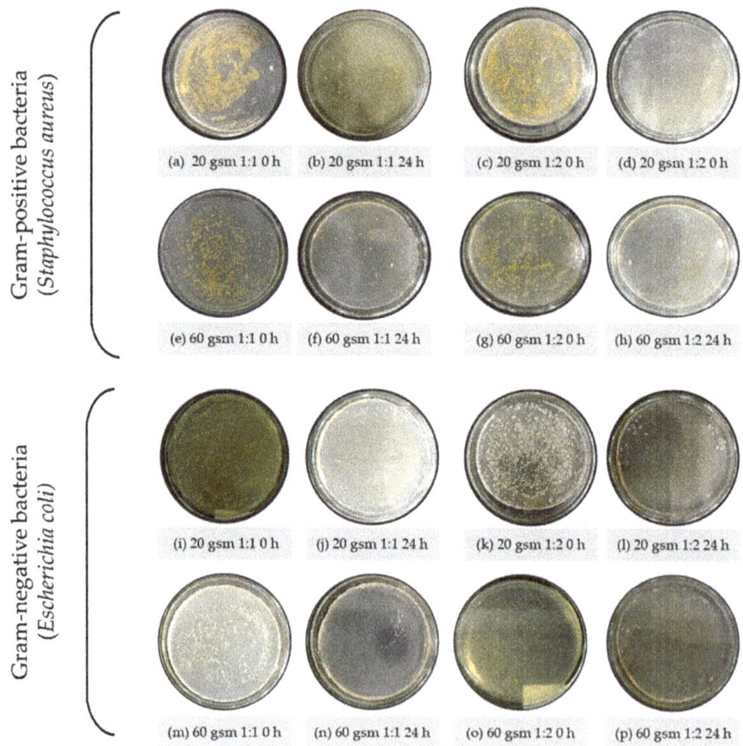

Appendix A Antimicrobial Tests (AATCC 100) for 20 and 60 gsm Fabrics (*Moringa oleifera*)

Appendix A.1. GC-MS Analysis

Chemical Constituents, Phytoconstituents and Compounds and Present in *Moringa oleifera*, *Aegle marmelos*, and Murraya koengii (curry leaf), and Coconut oil Using GC-MS Analysis.

Appendix A.1.1. *Moringa oleifera*

GC-MS analysis of *Moringa oleifera* showed sixteen chemical compounds from the leaf methanolic extract and they are as follows: 9-octadecenoic acid (20.89%), L-(+)-ascorbic acid-2,6- dihexadecanoate (19.66%), 14–methyl-8-hexadecenal (8.11%), 4- hydroxyl-4-methyl-2-pentanone (7.01%), 3-ethyl-2, 4-dimethylpentane (6.14%), phytol (4.24%), octadecamethyl-cyclononasiloxane (1.23%), 1, 2-benzene dicarboxylic acid (2.46%), 3, 4-epoxyethanone comprising (1.78%), N-(-1-methylethyllidene)-benzene ethanamine (1.54%), 4, 8, 12, 16-tetramethylheptadecan-4-olide (2.77%), 3-5-bis (1, 1-dimethylethyl)-phenol (2.55%), 1-hexadecanol (1.23%), 3, 7, 11, 15-tetramethyl-2 hexadecene-1-ol (1.17%), hexadecanoic acid (2.03%) and 1, 2, 3-propanetriyl ester-9 octadecenoic acid (1.23%), [58].

Five chemical constituents were identified in methanolic seed extract, and they are oleic acid (84%), L-(+)-ascorbic acid- 2, 6-dihexadecanoate (9.80%), 9-octadecenoic acid (1.88%), methyl ester-hexadecanoic acid (1.31%) and 9-octadecenamide (0.78%). These chemical constituents and bioactive compounds present in *Moringa oleifera* confirmed the medicinal values [58].

In addition, different types of active phytoconstituents like alkaloids, protein, quinine, saponins, flavonoids, tannin, steroids, glycosides, fixed oil and fats were present in *Moringa oleifera*. Furthermore, various other constituents were also present niazinin A, niazinin B and niazimicin A, niaziminin B [48]. Leaf extracts of *Moringa oleifera* were reported to inhibit the growth of *Pseudomonas aeruginosa* and *Staphylococcus aureus* [65]. These phytoconstituents and compounds enable the inhibition of microbial growth.

Appendix A.1.2. *Aegle marmelos*

GC-MS analysis of *Aegle marmelos* using methanolic extracts revealed twenty compounds of which Stigmasterol (14.49%), hexadecenoic acid (13.20%) and tetradecanoic acid (2.35%) were identified. These are fatty acids and provide antimicrobial resistance. The leaf extracts with methanol solvents showed high antimicrobial activity against *Escherichia coli*, *Klebsiella sp*. *Enterococcus sp*, and *Proteus mirabilis* [49].

Appendix A.1.3. *Murraya koengii* (Curry Leaf)

Curry leaf is widely consumed in Indian cuisine for its distinct aroma and medicinal value and used as in ayurvedic medicine for treating a variety of diseases [66]. GC-MS chromatogram of the methanolic extract of *Murraya koenjii* showed five compounds: α.-Caryophyllene, 2-phenyl-4-quinolinecarboxamide, Phenanthrene, 10H-Phenoxaphosphine, and 1,5-Diformyl-2,6-Dimethoxy-Anthracene [66]. These components enable antimicrobial resistance.

Appendix A.1.4. Coconut oil

Coconut oil is extracted from the dried kernel of the coconuts, widely grown in tropical countries, produces a mild yellow colour oil with 92% saturated fatty acids [50,67]. It is a vegetable oil with a high percentage of fatty acid (lauric acid 48.3% and 18.4% myristic acid) [68]. GC-MS analysis of coconut oil revealed various fatty acids, among them capric acid, lauric acid, myristic acid and palmitic acid, were in high proportions. It is worth mentioning that from the analysis that coconut oil has higher saturated fatty acid (87.2%) and lower unsaturated fatty acid (8.4%), which is mainly composed of oleic acid (5.5%) [68]. Saturated fatty acids have health benefits, including being antimicrobial and increasing body immunity [67]. Both lauric acid [69] and oleic acid [70] have many health benefits. Lauric acid derived from coconut oil has shown resistance to a wide range of Gram-positive and Gram-negative bacteria [69]. Monolaurin is an antimicrobial agen derived from lauric acid and glycerin [71]. Lauric acid is associated with treating a rang of skin conditions—acne, [72] and psoriasis [73].

Appendix B.

Tensile strength elongation.

Table A1. Tensile strength-elongation values for 20 and 60 gsm fabrics treated with nano-emulsion 1 and 2 from padded and exhaust method.

I	Breaking Extension (%) *Moringa oleifera*							
	20 gsm Padding		60 gsm Padding		20 gsm Exhaust		60 gsm Exhaust	
	Warp	Weft	Warp	Weft	Warp	Weft	Warp	Weft
Control	7.81 ± 0.00	27.33 ± 0.26	15.68 ± 0.14	14.675 ± 0.23	7.81 ± 0.00	27.325 ± 0.26	15.68 ± 0.14	14.68 ± 0.23
1:0.5 ratio	8.03 ± 0.21	29.25 ± 0.16	16.31 ± 0.24	12.51 ± 0.00	7.19 ± 0.49	26.23 ± 1.86	18.51 ± 1.09	15.31 ± 0.18
1:1 ratio	8.18 ± 0.00	26.34 ± 1.70	17.5 ± 0.97	12.49 ± 0.49	7.68 ± 0.09	27.71 ± 0.42	18.68 ± 0.52	11.71 ± 1.70
1:2 ratio	29.18 ± 1.15	20.41 ± 0.54	15.01 ± 1.34	11.33 ± 0.85	8.57 ± 1.49	25.51 ± 5.28	18.00 ± 0.64	13.55 ± 1.27
II	Breaking Extension (%) *Aegle marmelos*							
	20 gsm Padding		60 gsm Padding		20 gsm Exhaust		60 gsm Exhaust	
Control	7.81 ± 0.00	27.33 ± 0.26	15.68 ± 0.14	14.68 ± 0.23	7.81 ± 0.00	27.33 ± 0.26	15.68 ± 0.14	14.68 ± 0.23
1:0.5 ratio	7.94 ± 0.00	20.82 ± 0.00	14.44 ± 0.00	13.02 ± 0.00	6.07 ± 0.72	23.70 ± 0.23	15.23 ± 0.25	14.33 ± 0.06
1:1 ratio	6.62 ± 0.47	24.07 ± 0.72	17.78 ± 0.03	12.57 ± 0.26	8.08 ± 0.76	23.71 ± 0.69	15.44 ± 0.06	14.45 ± 0.11
1:2 ratio	8.34 ± 0.00	21.83 ± 0.00	17.73 ± 0.00	13.22 ± 0.00	7.45 ± 0.26	27.80 ± 3.12	15.75 ± 0.07	14.11 ± 0.02

References

1. Lim, S.; Hudson, S. Application of a fiber-reactive chitosan derivative to cotton fabric as an antimicrobial textile finish. *Carbohyd. Polym.* **2004**, *56*, 227–234. [CrossRef]
2. Neely, A.; Maley, M.P. Survival of enterococci and staphylococci on hospital fabrics and plastics. *J. Clin. Microbiol.* **2000**, *3* 724–726. [CrossRef] [PubMed]
3. Mitchell, A.; Spencer, M.; Edmiston, C. Role of healthcare apparel and other healthcare textiles in the transmission of pathogen A review of the literature. *J. Hosp. Infect.* **2015**, *90*, 285–292. [CrossRef] [PubMed]
4. Morais, D.S.; Guedes, R.M.; Lopes, M.A. Antimicrobial approaches for textiles: From research to market. *Materials* **2016**, *9*, 49 [CrossRef]
5. Pinho, E.; Henriques, M.; Oliveira, R.; Dias, A.; Soares, G. 2010 Development of biofunctional textiles by the application o resveratrol to cotton, bamboo, and silk. *Fibers Polym.* **2016**, *11*, 271–276. [CrossRef]
6. Massella, D.; Argenziano, M.; Ferri, A.; Guan, J.; Giraud, S.; Cavalli, R.; Barresi, A.; Salaün, F. Bio-functional textiles: Combi ing pharmaceutical nanocarriers with fibrous materials for innovative dermatological therapies. *Pharmaceutics* **2019**, *11*, 40 [CrossRef]
7. Hipler, U.C.; Elsner, P. Biofunctional Textiles and the Skin. In *Current Problems in Dermatology*, 1st ed.; Itin, P., Jemec, G.B.E., Eds Karger: Basel, Switzerland, 2006; Volume 33.
8. Carosio, F.; Cuttica, D.; di Blasio, A.; Alongi, J.; Malucelli, G. Layer by layer assembly of flame retardant thinfilms on closed ce PET foams: Efficiency of ammonium polyphosphate versus DNA. *Polym. Degrad. Stab.* **2015**, *113*, 189–196. [CrossRef]
9. Ferri, A.; Kumari, N.; Peila, R.; Barresi, A.A. Production of menthol-loaded nanoparticles by solvent displacement. *Can. J. Chem Eng.* **2017**, *95*, 1690–1706. [CrossRef]
10. Ghayempour, S.; Montazer, M. Tragacanth Nanocapsules containing chamomile extract prepared through sono-assisted W/O/W microemulsion and UV cured on cotton fabric. *Carbohydr. Polym.* **2017**, *170*, 234–240. [CrossRef]
11. Massella, D.; Ancona, A.; Garino, N.; Cauda, V.; Guan, J.; Salaun, F.; Barresi, A.A.; Ferri, A. Preparation of bio-functional textiles b surface functionalisation of cellulose fabrics with caffeine loaded nanoparticles. *IOP Conf. Ser. Mater. Sci. Eng.* **2018**, *460*, 01204 [CrossRef]
12. Massella, D.; Celasco, E.; Salaün, F.; Ferri, A.; Barresi, A. Overcoming the limits of flash nanoprecipitation: Effective loading o hydrophilic drug into polymeric nanoparticles with controlled structure. *Polymers* **2018**, *10*, 1092. [CrossRef]
13. Rivero, P.; Urrutia, A.; Goicoechea, J.; Arregui, F. Nanomaterials for functional textiles and fibers. *Nanoscale Res. Lett.* **2015**, *10*, 50 [CrossRef]
14. Silva, A.; Silvestre, A.; Freire, C.; Vilela, C. Modification of textiles for functional applications. In *Fundamentals of Natural Fibre and Textiles*; Textile Institute Book series; Elsevier: Amsterdam, The Netherlands, 2021; pp. 303–365.
15. Gao, Y.; Cranston, R. Recent advances in antimicrobial treatments of textiles. *Text. Res. J.* **2008**, *78*, 60–72.

16. Ristić, T.; Zemljič, L.F.; Novak, M.; Kunčič, M.K. Antimicrobial efficiency of functionalised cellulose fibres as potential medical textiles. In *Science against Microbial Pathogens: Communicating Current Research and Technological Advances*; Méndez-Vilas, A., Ed.; Formatex Research Center: Badajoz, Spain, 2011.
17. Glazer, A.N.; Nikaido, H. *Microbial Biotechnology: Fundamentals of Applied Microbiology*; Cambridge University Press: Cambridge, UK, 2007.
18. Sun, G. Antibacterial textile materials for medical applications. In *Functional Textiles for Improved Performance, Protection and Health*; Woodhead Publishing: Sawston, UK, 2011; pp. 360–375.
19. Hegstad, K.; Langsrud, S.; Lunestad, B.; Scheie, A.; Sunde, M.; Yazdankhah, S. Does the wide use of quaternary ammonium compounds enhance the selection and spread of antimicrobial resistance and thus threaten our health? *Microb. Drug Resist.* **2010**, *16*, 91–104. [CrossRef]
20. Kramer, A.; Guggenbichler, P.; Heldt, P.; Jünger, M.; Ladwig, A.; Thierbach, H.; Weber, U.; Daeschlein, G. Hygienic relevance and risk assessment of antimicrobial-impregnated textiles. *Curr. Probl. Dermatol.* **2006**, *33*, 78–109.
21. Windler, L.; Height, M.; Nowack, B. Comparative evaluation of antimicrobials for textile applications. *Environ. Int.* **2013**, *53*, 62–73. [CrossRef]
22. Islam, S.-U.; Shahid, M.; Mohammad, F. Perspectives for natural product-based agents derived from industrial plants in textile applications—A review. *J. Clean. Prod.* **2013**, *57*, 2–18. [CrossRef]
23. Upadhyay, A.; Upadhyaya, I.; Kollanoor-Johny, A.; Venkitanarayanan, K. Combating pathogenic microorganisms using plant-derived antimicrobials: A minireview of the mechanistic basis. *BioMed Res. Int.* **2014**, *2014*, 761741. [CrossRef]
24. Savoia, D. Plant-derived antimicrobial compounds: Alternatives to antibiotics. *Future Microbiol.* **2012**, *7*, 979–990. [CrossRef]
25. Sheikh, J.; Bramhecha, I. Enzymes for green chemical processing of cotton. In *The Impact and Prospects of Green Chemistry for Textile Technology*; ul-Islam, S., Butola, B.S., Eds.; Woodhead Publishing: Sawston, UK, 2019; pp. 135–160.
26. Eid, B.; Ibrahim, N. Recent developments in sustainable finishing of cellulosic textiles employing biotechnology. *J. Clean. Prod.* **2021**, *284*, 124701. [CrossRef]
27. Purwar, R.; Joshi, M. Recent developments in antimicrobial finishing of textiles. *AATCC Rev.* **2004**, *4*, 22–26.
28. Joshi, M.; Wazed, S.; Purwar, R.; Rajendran, S. Eco-friendly antimicrobial finishing of textiles using bioactive agents on natural products. *Indian J. Fibre Text. Res.* **2009**, *34*, 295–304.
29. Singh, R.; Jain, A.; Panwar, S.; Gupta, D.; Kharea, S.K. Antibacterial activity of some natural dyes. *Dyes Pigment* **2005**, *66*, 99–102. [CrossRef]
30. Scalbert, A. Antimicrobial properties of tannins. *Phytochemistry* **1991**, *30*, 3875–3883. [CrossRef]
31. Vastrad, J.V.; Goudar, G.; Byadgi, S.A. Characterisation of phenolic compounds in eucalyptus globulus and cymbopogan citratus leaf extracts. *Bioscan* **2016**, *11*, 2153–2156.
32. Jaswal, P.; Preet, A.; Simran; Goel, G. Antimicrobial activity of herbal treated cotton fabric. *Int. Res. J. Eng. Technol.* **2017**, *4*, 39–43.
33. Ijarotimi, O.; Adeoti, O.; Ariyo, O. Comparative study on nutrient composition, phytochemical, and functional characteristics of raw, germinated, and fermented *Moringa oleifera* seed flour. *Food Sci. Nutr.* **2013**, *1*, 452–463. [CrossRef]
34. Gopalakrishnan, L.; Doriya, K.; Kumar, D.S. *Moringa oleifera*: A review on nutritive importance and its medicinal application. *Food Sci. Hum. Wellness* **2016**, *5*, 49–56. [CrossRef]
35. Viera, G.; Mourão, J.; Ângelo, Â.; Costa, R.; Vieira, R. Antibacterial effect (in vitro) of *Moringa oleifera* and annona muricata against gram positive and gram-negative bacteria. *Rev. Inst. Med. Trop. São Paulo* **2010**, *52*, 129–132. [CrossRef]
36. Rockwood, J.L.; Anderson, B.G.; Casamatta, D.A. Potential uses of *Moringa oleifera* and an examination of antibiotic efficacy conferred by M oleiefera see and leaf extracts using crude extraction methods available to underserved indigenous population. *Int. J. Phototherapy Res.* **2013**, *3*, 61–71.
37. Atef, N.; Shanab, S.; Negm, S.; Abbas, Y. Evaluation of antimicrobial activity of some plant extracts against antibiotic susceptible and resistant bacterial strains causing wound infection. *Bull. Natl. Res. Cent.* **2019**, *43*, 144. [CrossRef]
38. Maity, P.; Hansda, D.; Bandyopadhyay, U.; Mishra, D.K. Biological activities of crude extracts and chemical constituents of Bael, *Aegle marmelos* (L.) Corr. *Indian J. Exp. Biol.* **2009**, *47*, 849–861. [PubMed]
39. Katayama, T.; Nagai, I. Chemical significance of the volatile components of spices in the food preservative viewpoint-IV. *Nippon Suisan Gakkaishi* **1960**, *26*, 29–32. [CrossRef]
40. Duke, J.A. *Handbook of Biologically Active Phytochemicals and Their Activities*; CRC Press: Boca Raton, FL, USA, 1992.
41. Pub Chem. National Library of Medicine. 2021. Available online: https://pubchem.ncbi.nlm.nih.gov/ (accessed on 17 May 2021).
42. George, M.; Venkataraman, P.R.; Pandalai, K.M. Investigations on plant antibiotics, part II A search ofr antibiotic substance in some Indian medicinal plants. *J. Sci. Ind. Res.* **1947**, *6B*, 42.
43. Joshi, C.G.; Magar, N.G. Antibiotic activity of some Indian medicinal plants. *J. Sci. Ind. Res.* **1952**, *11B*, 261.
44. Pitre, S.; Srivastava, S.K. Pharmacological, microbiological and phytochemical studies on the root of *Aegle marmelos*. *Fitoterapia* **1987**, *58*, 194. [CrossRef]
45. Pandey, D.K.; Asthana, A.; Tripathi, N.N.; Dixit, S.N. Volatile plant products vis-à-vis potato pathogenic bacteria. *Ind. Perf.* **1981**, *25*, 10–14.
46. Lambert, J.; Shurvell, H.; Lightner, D.; Cooks, R. *Introduction to Organic Spectroscopy*; Macmillan: New York, NY, USA, 1987.

47. Peter Wuelfing, W.; Kosuda, K.; Templeton, A.; Harman, A.; Mowery, M.; Reed, R. Polysorbate 80 UV/vis spectral and chromatographic characteristics—Defining boundary conditions for use of the surfactant in dissolution analysis. *J. Pharm. Biomed. Anal.* **2006**, *41*, 774–782. [CrossRef]
48. Dhongade, H.; Paikra, B.; Gidwani, B. Phytochemistry and pharmacology of *Moringa oleifera* lam. *J. Pharmacopunct.* **2017**, *20*, 194–200. [CrossRef]
49. Murugesan, K.; Ezhilarasu, A.; Gajendiran, K. Phytochemical screening, antioxidant and antimicrobial activity of *Aegle marmelos* from Kolli Hills. *Int. J. Interdiscip. Res. Rev.* **2013**, *1*, 10–15.
50. Yousefi, M.; Nateghi, L.; Rezaee, K. Investigation of physicochemical properties, fatty acids profile and sterol content in malaysian coconut and palm oil. *Ann. Biol. Res.* **2013**, *4*, 214–219.
51. AATCC 100. *Assessment of Antibacterial Finishes on Textile Materials: Developed by American Association of Textile Chemists and Colorists*; American Association of Textile Chemists and Colorists: Research Triangle Park, NC, USA, 2019.
52. AATCC 147. *Antibacterial Activity Assessment of Textile Materials: Parallel Streak Method Developed by American Association of Textile Chemists and Colorists*; American Association of Textile Chemists and Colorists: Research Triangle, NC, USA, 2016.
53. AATCC 30-III. *Assessment of Antifungal Activity on Textile Materials: Agar Plate Developed by American Association of Textile Chemists and Colorists*; American Association of Textile Chemists and Colorists: Research Triangle, NC, USA, 2013.
54. ASTM 5035-11. *Standard Test Method for Breaking Force and Elongation of Textile Fabrics (Strip Method) ASTM (American Society for Testing and Materials) International*; ASTM: West Conshohocken, PA, USA, 2019.
55. IS 15370. *Indian Standard for Domestic Washing and Drying Procedure for Textile Testing*; Standard identical to ISO 6330: New Delhi, India, 2000/2005.
56. Korgaonkar, S.; Sayed, U. Nano herbal oils application on nonwovens. *Int. J. Pharm. Res.* **2021**, *13*, 1965–1970.
57. Rajendran, R.; Radhai, R.; Kotresh, T.; Csiszar, E. Development of antimicrobial cotton fabrics using herb loaded nanoparticles. *Carbohydr. Polym.* **2013**, *91*, 613–617. [CrossRef]
58. Aja, P.; Nwachukwu, N.; Ibiam, U.; Igwenyi, I.; Offor, C.; Orji, U. Chemical constituents of *Moringa oleifera* leaves and seeds from Abakaliki, Nigeria. *Am. J. Phytomed. Clin. Ther.* **2014**, *2*, 310–321.
59. Abidi, N.; Cabrales, L.; Haigler, C. Changes in the cell wall and cellulose content of developing cotton fibers investigated by FTIR spectroscopy. *Carbohydr. Polym.* **2014**, *100*, 9–16. [CrossRef]
60. Innovation in Textiles. Antimicrobial Textiles Market to Reach $12.3 billion. 2019. Available online: https://www.innovationintextiles.com/antimicrobial-textiles-market-to-reach-123-billion/ (accessed on 5 August 2020).
61. Venkatraman, P.D.; Sayed, U.; Parte, S.; Korgaonkar, S. Consumer perception of environmentally friendly antimicrobial textiles: A case study from India. *Int. J. Adv. Sci. Eng.* **2021**, *10*, 1782–1793.
62. Muhuha, A.W.; Kang'ethe, S.K.; Kirira, P.G. Antimicrobial activity of *Moringa oleifera*, aloe vera and warbugia ugandensis on multidrug resistant escherichia coli, pseudomonas aeruginosa and staphylococcus aureus. *Int. J. Antimicrob. Agents* **2018**, *4*, 168.
63. Dike-Ndudim, J.N.; Anyanwu, G.O.; Egbuobi, R.C.; Okorie, H.M.; Udujih, H.I.; Nwosu, D.C.; Okolie, N.J.C. Anti-bacterial and phytochemical potential of *Moringa oleifera* leaf extracts on some wound and enteric pathogenic bacteria. *Eur. J. Bot. Plant Sci. Phytol.* **2016**, *3*, 50–60.
64. Dzotam, J.K.; Touani, F.K.; Kuete, V. Antibacterial and antibiotic-modifying activities of three food plants (Xanthosoma mafaffa Lam., *Moringa oleifera* (L.) schott and passiflora edulis sims) against multidrug-resistant (MDR) gram-negative bacteria. *BMC Complement Altern. Med.* **2016**, *16*, 9. [CrossRef]
65. Caceres, A.; Cabrera, O.; Morales, O.; Mollinedo, P.; Mendia, P. Pharmacological properties of *Moringa oleifera*: Preliminary screening for antimicrobial activity. *J. Ethnopharmacol.* **1991**, *33*, 213–216. [CrossRef]
66. Balasubramanian, S.; Dama, G.; Surya Narayana, V.V.S.; Reddy, S. GC-MS analysis of the curry leaves (Murraya Koengii). *Glob. J. Biol. Agric. Health Sci.* **2014**, *3*, 3–10.
67. Gopala Krishna, A.G.; Raj, G.; Bhatnagar, A.S.; Prasanth Kumar, P.K.; Chandrashekar, P. Coconut oil: Chemistry, production and its applications—A review. *Indian Coconut J.* **2010**, *73*, 15–27.
68. Moigradean, D.; Poiana, M.-A.; Alda, L.-M.; Gogoasa, I. Quantitative identification of fatty acids from walnut and coconut oils using GC-MS method. *J. Agroaliment. Process. Technol.* **2013**, *19*, 459–463.
69. Dayrit, F. The properties of lauric acid and their significance in coconut oil. *J. Am. Oil Chem. Soc.* **2014**, *92*, 1–15. [CrossRef]
70. Sales-Campos, H.; Reis de Souza, P.; Crema Peghini, B.; Santana da Silva, J.; Ribeiro Cardoso, C. An overview of the modulatory effects of oleic acid in health and disease. *Mini Rev. Med. Chem.* **2013**, *13*, 201–210.
71. Monolaurin. What is Monolaurin? 2021. Available online: https://www.healthline.com/health/monolaurin (accessed on 12 July 2021).
72. Nakatsuji, T.; Kao, M.; Fang, J.; Zouboulis, C.; Zhang, L.; Gallo, R.; Huang, C. Antimicrobial property of lauric acid against propionibacterium acnes: Its therapeutic potential for inflammatory acne vulgaris. *J. Investig. Dermatol.* **2009**, *129*, 2480–2488. [CrossRef]
73. Health Line. What is Lauric Acid? 2021. Available online: https://www.healthline.com/health/beauty-skin-care/what-is-lauric-acid#for-psoriasis (accessed on 12 July 2021).

MDPI
St. Alban-Anlage 66
4052 Basel
Switzerland
Tel. +41 61 683 77 34
Fax +41 61 302 89 18
www.mdpi.com

Coatings Editorial Office
E-mail: coatings@mdpi.com
www.mdpi.com/journal/coatings

www.ingramcontent.com/pod-product-compliance
Lightning Source LLC
Chambersburg PA
CBHW062105220926
51850CB00010B/2900
* 9 7 8 3 0 3 6 5 2 0 8 1 0 *